BEEKEEPING

A Beginner's Guide to Becoming a Successful
Backyard Beekeeper

(How to Start, Finance & Market a Beekeeping
Business)

Bobbie Byrd

Published by Oliver Leish

Bobbie Byrd

All Rights Reserved

Beekeeping: A Beginner's Guide to Becoming a Successful Backyard Beekeeper (How to Start, Finance & Market a Beekeeping Business)

ISBN 978-1-77485-186-9

Legal & Disclaimer

The information contained in this book is not designed to replace or take the place of any form of medicine or professional medical advice. The information in this book has been provided for educational and entertainment purposes only.

The information contained in this book has been compiled from sources deemed reliable, and it is accurate to the best of the Author's knowledge; however, the Author cannot guarantee its accuracy and validity and cannot be held liable for any errors or omissions. Changes are periodically made to this book. You must consult your doctor or get professional medical advice before using any of the suggested remedies, techniques, or information in this book.

Upon using the information contained in this book, you agree to hold harmless the Author from and against any damages, costs, and expenses, including any legal fees potentially resulting from the application of any of the information provided by this guide. This disclaimer applies to any damages or injury caused by the use and application, whether directly or indirectly, of any advice or information presented, whether for breach of contract, tort, negligence, personal injury, criminal intent, or under any other cause of action.

You agree to accept all risks of using the information presented inside this book. You need to consult a professional medical practitioner in order to ensure you are both able and healthy enough to participate in this program.

Table of Contents

Introduction

We all have heard of land flowing with honey and milk. It would be a great dream to have unlimited honey at your disposal.

Honey can have many health benefits, including fighting infections and weight loss. Honey is a wonderful product. There are very few negative aspects.

Honey is costly. Although you've searched around and tried to find the best honey without spending a lot, you still end up with a small jar of honey and a empty wallet. It is important to be careful about how much you spend on sweet substances. But you also know that there are better ways to do this.

How? It's impossible to go into the woods to find a beehive right now. It is difficult to find a honey-hive. You would also have to endure the stings from thousands of bees, who will all die for their hives.

You will have to do it in a different way.

You will need to take care of your own bees.

You think about it for a moment. How are you going to do that? Is it not difficult to keep bees healthy? How can you ensure they stay warm in winter and keep the hive occupied in summer? How do you get bees?

It's easy for you to feel overwhelmed by all of these questions. But don't worry. You have reached the right place if you're serious about your beekeeping and beekeeping career.

I will show you how to get started with beekeeping and how to keep it going.

This book will change the way you view bees and how you obtain your honey. You will be able to see the sweetness of this deal by the time you reach its end.

Let's get started.

Chapter 1: History of Beekeeping

Beekeeping dates back thousands of years. Rock paintings show that honey collection dates back to 13,000 BC. We know that the Egyptians kept honey bees from very old drawings of beekeepers in the Niuserre sun temple, which dates back to the 5th Dynasty (previously 2422 BC).

Drawing shows bee-keepers inhaling smoke as they take out honey-combs from the bee hives. Pots of honey were also found in the graves of King Tut and some other Pharaohs.

Cave drawings showing beekeeping from the Stone Age have also been discovered in Spain. Cave drawings dating back to 6000 B.C. Shows people harvesting honey from beehives.

The Tacuina Sanitatis is a 14th-century medieval handbook that shows bees and beehives. Bee keeping has been a tradition for thousands of years. It's been done in the same manner for thousands of years. Honey is so delicious, why shouldn't beekeeping be a part of our daily lives?

Beekeeping has been practiced on all continents, except Antarctica, and all societies. In the beginning, beekeepers were just able to raid the beehives, and then destroyed the hives.

Then, somewhere along the way, humans started creating beehives and domesticating bees. The first artificial bee hives were made from hollow logs, large pottery vessels and straw baskets.

There is evidence to suggest that an advanced and well-organized beekeeping system existed in ancient Israel around 3,000 years ago. Amihai Mazar, a Hebrew

University of Jerusalem archaeologist, discovered bee hive ruin ruins that date back to 900 BC. These ruins contained around 100 hives and more than 1,000,000 bees.

In the 16th century, beekeeping was introduced to America. John Harbison brought the first apiculture system to Pennsylvania in the 1800s. Due to food shortages and high prices of food, beekeeping became very popular during the Great Depression and World War I.

Sears and Roebuck catalogues were used to order bee hive equipment. The wild bees were then moved from the trees to the new hives. In those days, people had to make do with what they had. Beekeeping was a huge help.

After World War II, beekeeping was not popular. However, the Colony Collapse Disease (bees disappearing from hives across the U.S.) has brought attention to honeybees and more people are interested in beekeeping again.

This is great news for honey bees and our food system, as without them pollinating our crops, over 2/3s wouldn't grow.

The renewed interest in backyard beekeeping isn't just for rural areas. Tokyo, New York City, London, Paris, London, New York City, and Washington D.C. are all encouraging urban beekeeping. These are the most important urban beekeeping areas worldwide.

Scientists are still trying to find out what causes Colony Collapse Disorder. It is vital that people take up apiculture by thousands. Our crops, flowers, apples trees and other plants depend on bees for pollination.

Chapter 2: A Short History of Beekeeping

Honey collection dates back to approximately 13,000 BC. North African pots depicting beekeeping date back 9,000 years. Beekeeping is one of the oldest forms for food production. Bees are the only insect that can produce food for humans. To harvest their honey, Ancient Egyptians used simple hives and smokers. Artifacts and writings also show that beekeeping was practiced in Ancient Greece, Ancient China, and other countries.

Primitive beekeepers were part of early cultures that relied heavily on hunting and gathering. However, when harvest time came, they destroyed the entire hive to collect honey. This also killed the queen and all the larvae. The colony was effectively destroyed and they had to start again. This was also not the most sanitary way to collect honey. The 18th and 19th century saw a revolution in beekeeping. In Europe, moveable hives were invented. This allowed beekeeping technology rapid advancements. People also began to live a more nomadic lifestyle. The destruction of the hive was no longer a necessity.

The United States is thought to have been founded in 1600 by beekeepers. The Western honey bees were imported from Europe because they aren't native to the United States. They were widespread from the Atlantic Ocean to Mississippi River by 1800.

John Harbison brought beekeeping to America's west coast in the 1860s. It was done eagerly so people could enjoy their honey and wax. It is believed that natural swarms originated in California and traveled north to Washington before reaching the mountains.

Prior to the 1980s, most beekeepers were farmers or relatives. It was difficult to find equipment that could be used to keep hives on land other than farmland. It is now easier than ever to get started in beekeeping.

The tracheal and varroa mites are two of the most recent challenges facing American beekeepers. These mites were introduced in 1980. Small hive beetles were also an issue in the 1990s. These pests can be controlled with chemical and natural methods, but hybrid bees are able to resist them.

Chapter 3: Building your own Hive

A Top Bar Hive or Honey Cow

There are two options: you can either buy pre-built hives or build your own. There are many types of hives. The Honey Cow or Top Bar hive is the most popular.

These are the instructions to help you build your Honey Cow.

Materials

55-gallon food-grade plastic barrel - This will make two beehives

1 x 2 lumber - 22 feet

1 1/2 x1 lumber - 46 feet

2 x 4 lumber, 2 x 8 feet lengths

Tin - 3 feet. x 4 feet.

1 1/2 inch wood screws - 20

2-inch wood screws - 10

1/2 inch screws - 8

Tie wire or bungee cord

45 feet of thin molding

Tools

A circular or jigsaw saw?

Drill

Tape measure

Marker

Tiny snips

Step 1 - The Barrel

Reduce the barrel in half lengthwise. Make sure that each half of the barrel has a bung hole

You don't know what may have been inside the barrel before you clean it. You should try to find a food-grade container, as they are less likely to have contained chemicals that could be deadly.

Place the barrel halves flat on the ground. Turn the cut side up. This is how the barrel will be positioned for the hive.

The plastic rim can be found at the end of the barrel, where it was once complete.

Use beeswax to smoothen the entire interior. This will remove any unpleasant odors and make the interior more appealing to bees. If you prefer, you can add a few drops of lemongrass to your bees.

Step 2 - The Frame

Measure the barrel's width and length using your tape measure. Then cut the 1x2 wood for the frame. You should make sure that the length is 1 inch more than the actual measurement. For example, if your barrel measures 36 by 24 inches by 24 inches, cut two pieces of 25-inch and two pieces of 37-inch. You can attach the two pieces together using screws by adding an inch.

Attach the frame to the wall and screw the pieces together.

Place the barrel in the frame and screw it into place.

The 2 x4 wood should be cut into 40-inch pieces - these are for the legs.

Attach each leg to the barrel sideways and screw the frame to it. To make sure the legs are strong and sturdy, attach them with several screws.

Step 3 - Top Bars

Cut 23 pieces of 24 inch-long wood from the 1 1/2 x 1. These will be the bars the honeybees use to attach the honeycomb. You will however

You will need to provide a little guidance to ensure that the combs are straight. This can be done in a variety of ways.

a) Cut the molding in 20-inch pieces. Screw one piece to each top bar. When the bar is in place, the molding will face down into the barrel. Apply beeswax to the molding

OR

b) Wrap a length of twine with wax, and attach it to the top bars the same way you attached the molding.

OR

c) Cut a narrow groove in your top bar. Then, pour molten beeswax inside. The groove should be about 1/4 inch in width and the end should not exceed one inch.

Step 4 - The Roof

To make a frame that fits around the barrel's frame, use the 1x2 lumber. To make the roof frame, cut two pieces of 27 1/2 inch and two of 37 inch frames.

Fix the tin to your frame and ensure an equal overhang.

Attach the overhang to the frame sides by bending it down.

Take off all bits hanging below the frame

Install the roof on the barrel frame's top

Then wrap the bungee cords around everything, including the barrel and roof. Join them together. Tie wire can be used to secure the roof to the barrel.

Step 5 - It's time for the Bees

The hive is now ready for you to start the introduction of bees. You can either purchase a package with the queen, her workers, and drones, or you can capture your own swarm.

Don't think of bees only as individuals. The whole animal is the hive. If the conditions are right, the entire colony can reproduce. If they have enough food and the space is empty, they will make another queen.

They will be full of food, and will be busy looking for a home. Protective clothing is essential when dealing with a swarm. Even docile bees can sting you. You can find a swarm on a branch by placing a box under it and shaking the branch. You can then take the box to your hive and they will fall into a container. The rest will be up to the bees.

How to check up on your hive

It is best to let your bees do their thing. They are usually very knowledgeable about what they are doing. A beekeeper's role is to monitor and oversee the bees. You should check on your bees once a week unless something is very serious. You will want to ensure that your bees are healthy and producing eggs. Before you take any corrective actions, it is a good idea to speak with an experienced beekeeper if the bees seem slow or if you suspect they are sick. Even if you smoke, each hive-check will take a few minutes before the bees start to get excited. A digital camera allows you to take your time and look for signs of trouble. You should also expect that the bees will take several days to settle down completely after each hive check.

Take the honey when it is time to harvest it. This is the best time for the bees to get in a good mood. You can take what you need and do it quickly and gently. Then, move it to the enclosed place. Simply remove the honeycomb from the bar that it is attached to and scrape it or cut it off. This can be done with a centrifugal extractor. The extraction leaves enough comb to reduce the work required by the bees to rebuild the comb. You shouldn't expect to receive too much honey in the first year of your hive. The bees' primary concern is stability and maintaining its population. Many beekeepers don't take any honey from their first year. This gives beekeepers the best chance to establish a strong colony that will produce plenty of honey in future.

Here are some tips for keeping your bees healthy.

Remain calm. Avoid sudden movements. Try to be as relaxed and confident as possible.

You can wait for a sunny day to do your maintenance and then go outside during the afternoon. It is a good rule to not open a hive below 60 degrees. This will let the heat escape and could chill the eggs.

Avoid wearing dark or brown clothing. It is best to wear light colors. Brown and other dark colors can lead to confusion with bears or similar criminals. Wear gauzey or loosely-woven clothing. The barbs from bees' legs can get caught in the fabric and panic them.

The process of caring for bees can seem simple but it is not difficult. Once you are familiar with the process, and feel more comfortable with your bees it will become enjoyable to interact with these fascinating insects.

The Hive through all Seasons

Because bees don't work according to a schedule, there is no set beekeeping year. You can work your hive according to the seasons. This means that how and when you do it will change from one year to another. There is a plan that you can follow month by month. You just need to adapt it as you go. You need to complete certain tasks to ensure your honey bees are happy and healthy. You will become more aware of the conditions and be able to determine what task is needed.

January

It is a great time to begin building your hives or other equipment long before you actually need it. It is much easier to start building your hives now than to scramble for materials after you receive your package or when you catch the perfect swarm.

If your hives are already set up, January can be a very hands-off month. January is a great month to set up your hives.

February

February is an important month in beekeeping. You need to make sure that your hives are stocked with enough honey. This is especially important in warmer climates where bees are more active than staying in clusters to stay warm. However, bees in colder climates need sufficient honey to keep them warm and their hives warm.

To verify, lift one end of your hive and feel how heavy it is. It should be between 50 and 70 lbs. Honey is a good choice for feeding them. Keepers often keep a few frames worth of honey in reserve so they have some to put into their hives when needed. You can feed them honey or something else, such as fondant or a candyboard.

March

March can be an unusual month for beekeepers. The risk of your bees dying from starvation is much greater than at other times. It is important to constantly check the weight of your hives and feed them when necessary.

Bees are constantly on the move, so they need to be adequately fed. If possible, wait until it is warmer to open the hives. The mites can be treated in March before the nectar flows.

March is the best time to focus on stopping swarms. This is done by checking board. This is when you alternate honeycomb with empty brood frames on top of your brood chambers. This will give the bees the illusion of having space. The brood chamber is usually located in the middle of a hive. It's full of eggs and larvae and hasn't been touched. This area will be spread out by checkerboarding. You don't need any additional honey to feed your brood frames; just move an empty honeycomb in the place you have moved one.

April

Southern regions will see bees at full speed now. To make sure you get all the honey you want, you'll need to add additional boxes to your hives. The start of pollen production will take longer in cooler regions.

You can inspect the hives to see what's going on. You will need to spend some time to learn how to read the bees' behavior and understand what is happening. You should inspect the honeycombs to see if eggs are being laid or if pollen has been introduced into the hive. Inspect the hive for signs of disease and pests.

May

Northern beehives are at their best during this time of year. There should be plenty of activity at the entrance. The hive should smell of honey and bee wax. You should be able to see the nectar and pollen stores when you open the hive. You should ensure that there is a consistent pattern of brood chamber laying. It is a good time for you to watch out for signs of swarming. To create the illusion of more space, you can use the March checkerboard method. You can identify the queen cells, and then you can add a super to the hive when the nectar starts flowing.

The bees in the southern regions are busy during May, so you might start to notice capped honey in the frames. Keep an eye on your beehives to ensure that the queen is still alive and laying eggs.

June

The bees are in full swing due to the good weather. You should check your hives to ensure that they have a good laying pattern, and that the queen is well. In the northern areas, you can add supers, while in the southern areas, beekeepers begin to remove supers and bring in their first harvest.

July

The northern regions should be able to cap the nectar. The reducer should be removed at the entrance of the hive to allow the bees to cool down the hive.

The bees in the south have stopped producing honey, so it is time to split the hive to allow the bees to continue building their honey supply before winter sets in.

You can put the queen and half the colony in a new hive and leave half the colony without a queen. This will ensure that there is no brood for 42 consecutive days. It will also help you solve any Varromite problems.

This is the best month to start collecting sourwood honey if you have bees in the Appalachians. Others may not be able take advantage of the late nectar flows.

It is also a good idea to prepare your hives for winter. Take care of weak hives and get rid pests. Honey stores should also be monitored.

August

August is a good month to put back the entrance reducer to protect the bees from yellow jackets. You can close the hive for 24 hours if you have yellow jackets. Otherwise, leave the bees alone. The bees will be more hungry and active because there won't be much nectar. Make sure they have enough pollen to last them for the day and that honey is available. As a method of mite control, you can put a foundation frame for drone cells into your hive.

September

Northern beekeepers may be able to obtain late nectar from some flowers, but sugar syrup can be fed to the bees to ensure they have enough food for winter.

Once you have removed all supers, it is time to apply your chemical mite control. To avoid contamination, make sure to do this prior or after honey production.

Make sure the hive is sufficiently heavy. It should be 120 to 130 pounds if it has two deeps with a medium super inside.

October

No matter where they live, all beekeepers should inspect their hives for any pests and ensure that there is enough honey. You can either feed sugar syrup or use the honey reserves.

November

Keep feeding the sugar syrup to the bees as long as they continue to consume it. You can put mouse guards on your hives, and if you haven't, the entrance reducer.

December

It's time to prepare for the cold winter ahead. Stop mite treatment. Do not open the hives, except to remove the mite control strips.

December is the best time to enjoy the fruits of your and your bees' hard work. You can also catch up on any beekeeping magazines or books that are out, and check out new species that you might like to try next year.

Chapter 4: Common Bee Diseases

American Foulbrood

This is a serious bacterial infection that affects larvae and pupae. This is a contagious disease that affects bees, not humans. If left untreated, it can cause the death of all your bees and even your entire colony. This is the most severe bee disease. These are just a few symptoms.

After being capped, the infected larvae change from a healthy, pearly white whitish colour into dark brown or tan.

Dead brood caps sink inwards, becoming concave. They often have tiny holes and look perforated.

The capped brood's pattern is not compact but rather random and scattered. This is sometimes called a "shotgun" or "shotgun" pattern.

The cappings' surface may appear greasy or wet.

You should check if you see any of these symptoms. To confirm, insert a toothpick or matchstick into the brood and stir it. Then slowly remove the matchstick.

As you take out the matchstick, notice the material being pulled out of the cell. AFB kills brood and it will appear stringy. The rope will extend about 1/4 inch, then the material will reshape back into a rubber band. This test can be used for testing the presence of AFB and also to observe the pupae.

To confirm your suspicions, you should immediately contact a bee inspector. Your equipment and hives could be destroyed if your bees are diagnosed with AFB. AFB's sleeping pores can remain active for up 70 years.

Preventing the infection is possible by not buying any old equipment or using it. Also, you can treat your colonies with honey-recommended antibiotics in autumn and spring.

European Foulbrood (EFB)

This is a bacterial infection of larvae. This is a different disease than the AFB. Here, infected larvae are killed before they can be capped. These are some of the symptoms.

Spotty brood patterns are caused by many empty cells scattered among the capped brood. This is also known as a "shotgun" or "shotgun" design.

Infected larvae look like an inverted corkscrew. The infected larvae have a brown or light tan colour and a smooth, melted appearance. Remember that healthy larvae are typically bright white.

EFB causes almost all larvae to die in their cells, before they can be capped. It is usually easier to remove discolored larvae.

Capped cells can be perforated and sunken in a "toothpick test" described in the AFB. However, this will not produce a telltale ropy trail.

Although there might be a sour smell, it is not as bad as AFB.

Nosema

This is a common protozoan infection. It affects adult bees' intestines in a manner similar to dysentery in humans. It can lead to a decrease in honey production and hive weakness. It can decimate entire colonies of bees. After winter bees are kept in the hive, it becomes more apparent in spring.

The problem is usually that symptoms are often not apparent until it is too late. These are some of the symptoms.

If they do, the infected colonies will slowly build up during spring.

The entrance to the hive is where the bees are weakest.

The hive will look spotted, with streaks of brown faeces visible in the hive.

To reduce Nosema, a Hive site that has good ventilation and is close to a source of water can be used. Cold environment can encourage Nosema, so avoid damps. Your hive should be lit by sunlight.

You can increase air flow by making an upper entrance in winter. This reduces Nosema. To reduce the possibility of infection, make sure your queens or bees come from suppliers who use anti-biotics.

Chalkbrood

This is a very common disease that affects bee larvae. They are most common in the early spring when there is dampness. Chalkbrood is usually not very serious. Infected larvae turn white after a while and then turn black.

Chalkbrood can be identified by chalky carcasses found on the "front porch" of the hive. Worker bees may attempt to eliminate chalkbrood as soon as possible, but most often they will just drop the loads on the ground at the front door.

It is possible to give chalkbrood the wrong medication. This is very common. It is possible to give chalkbrood the wrong medication. This is because it is also effective for chill brood (which will be discussed later in the book). The colony can survive the disease on its own.

The only way to speed up the process is by helping to remove the carcasses from the ground surrounding the hive. The bees will have a much easier job cleaning up if they remove the frame that has most chalkbrood cells. It is worth replacing the queen to stop the spread of the fungus.

Sacbrood

This is a brood virus disease that is closely related to the common flu. Sacbrood does not pose a risk to your health. Sacbrood is not a serious disease. Infected larvae turn yellow and then brown. They look like they are in a water-filled bag, making it easier to get them out of a cell.

Sacbrood does not recommend any medical treatment at this time. You can reduce the disease quicker by removing the sacs using tweezers. Although the bees can do it themselves, an extra hand is helpful.

If you keep your bees away from stressors like mites, poor ventilation and crowded areas, they will have a better chance of staying healthy. You can feed them sugar syrup and pollen substitute in spring, and sugar syrup and syrup in autumn.

Stonebrood

Both the larvae as well as the pupae can be affected by this disease. This disease is rare because it rarely occurs. They transform the brood into mummified mummies. Mummies are typically solid and strong. They are not sponge-like or chalky, unlike chalkbrood. Brood can be covered with green fungal.

Stonebrood does not require any medical treatment. As worker bees remove most of the dead brood, the colony will recover on its own.

It is possible to speed things up by removing any mummies from the entrance and around your hive. You should also inspect the hive for infested frames and get rid of them.

Do not shake sick bees, as this could expose other healthy bees. This applies if you have more than 1 hive.

Chapter 5: Maximizing Honey Production

Honey is the main purpose of keeping bees. The yield of honey varies from one year to the next, and from colony to colony according to the availability and management within the apiary environment. Forage season is when the colonies are more active and honey is stored more.

There are many factors that affect the amount of honey per colony. These are the key factors.

Bee pasture availability

Management of the Hive

Honey storage and nesting

Hive population

Nutrition

There is plenty of space in the hive to allow for the expansion of the brood

The Queen's condition

The colony is free from pests

Predators

Experience of the beekeeper in both apiary as well as honey production

Weather

Swarming

Honey production will increase with colonies that have efficient queens and a high number of foraging honey bees. Good hive management skills, and favorable weather conditions are all key factors.

How to Get High Honey Yields

It is important to choose a place with nectar-producing plants to allow the bees to forage.

This includes guava, beans, cucumbers, sunflowers, eucalyptus and thistles as well as many other flowering herbs and shrubs.

It is important that the distance between the hive and the nectarous plants doesn't exceed 1 km. This is the best location for honey collection. The nectar-producing plants can still be planted in the area to let the bees eat it.

The bees would find it difficult to fly more than one kilometre to reach the plant from their hive. Honey production will increase if the plant is closer. Honey production is faster in hives located closer to the nectar source.

For big honey harvest, it is best to keep your hive 4km from any bee hives. This will allow for maximum honey production. It is unhealthy to have apiaries within close proximity of each other.

This can lead to overpopulation and low honey harvest. For maximum production, apiaries should have 3 to 4 hives on a 100 x 100 meter plot. This is influenced by the strength and attractiveness the crops.

To ensure maximum performance, the colony must be led by a virtuous queen. A queen can live up to two years, or more. However, you should know that egg laying decreases after each year, which reduces the colony's production capacity. It is recommended to replace the queen at the beginning of the second year.

To monitor and know the efficiency of the queen, you might need to keep checking her eggs as often as possible. This will allow you to determine if the queen is laying one cell at a time. If it isn't, then it could be an indication that the queen is old or not a queen at all. This means that you will need to replace the queen by a better one.

The colony will benefit from a young queen to increase honey production and help in the establishment of strong populations. You can also expect that some colonies will still produce high quantities of honey in their third year. The bees replace an old or fallen queen by creating supersedure cells near the centre of the honeycomb. This is done without the involvement of the bookkeeper.

If you see a few queen cells on your comb's face, you should not destroy them. They are supersedure cells that indicate an old or fallen queen. You should leave it as it is. It's better to let the bees replace the queen than you do it.

As colonies grow stronger, more honey will be produced. You can ensure that colonies are fully stimulated to brood activity six weeks before the blossom of major nectar-producing plants.

If colonies are able to build up their populations after or during the nectar flow, only a small amount of honey can be produced. Because they are a small foraging force, it is because they spend most of their time feeding their brood. It takes about 42 days for an egg to become a forager bee. A healthy colony should have at least 50,000 bees in the early dry seasons. Your colony should be ready for nectar flow from mid-September to November, if you are a beekeeper.

Brood can be seen extending over four bars or frames. Colonies must have combs to allow the nest to expand and for the queen to lay eggs.

Over-crowding can be caused by too many young bees and insufficient space for brood growth. Also, combs stuffed with honey and pollen may lead to over-crowding. To monitor the development of your bees, it is important to inspect the hive at least once a week. Regular observation will show you how to make more space for comb expansion.

As a beekeeper, it must be your top priority to increase the colony's population before the beginning of the nectar flow. Keep in mind that honey yields more honey if there are more bees. This means that honey from stronger beehives will yield a greater harvest.

Because the bees require water to keep the hive cool in hot weather, and to dilute honey, it is important to provide water closer to the hive. It will reduce the time required to gather nectar and maintain honey production by providing water near the hive.

Good hive management involves adding five bars at a time to allow the bees draw the combs. To do this, you can only start the colony with five bars. Then block the remaining bars with division boards. You can add the empty bars to the first set by moving the boards. Continue adding bars to the first set until you have enough combs.

This will allow you to have enough combs to withstand incoming nectar, and increase your large bee population. This will encourage foraging and decrease the early swarming.

You can inspect your hive periodically for honeycombs during nectar flow. When you are done, remove the entire capped combs and return them to your hive.

Queen excluder slows down the bees' speed and decreases honey storage in the combs. It should therefore be eliminated immediately.

Swarming reduces honey production because of the loss of large numbers of bees. This is due to the fact the old queen takes 30-60% of the bees from the parent colony. Swarming can be caused by the presence of a queen older than one year. This causes overpopulation, congestion, and population explosion.

If hive has too many queen cells, it can lead to congestion. When large numbers are created, the hive becomes congested. Large numbers of drones are then raised. These are indicators of swarming.

You must control swarming as a beekeeper. Below are some ways to control and prevent swarming.

Queen cells should be removed: It is recommended that you inspect the hive frequently during queen cell swarming season. You must immediately destroy queen cells if you see that they are being constructed. There is a greater likelihood of them hatching.

Re-queening: Colonies with younger queens are less likely to swarm, while colonies with older queens have a 30% chance of swarming. It is best to change a queen before the nectar flow period. Between mid-September to December and with a young queen, in order to lessen the tendency of swarm.

Space is essential for the bees to store honey and pollen throughout the season. If a colony grows too large or crowded, swarming is the only way to save it.

You can relieve the colony by moving a few broodcombs to another hive you want to strengthen.

After you have removed weak queens, it is possible to combine weak colonies with stronger ones that contain better queens for better results. Because weak colonies can't produce enough honey to sustain large numbers of people, this is why you should combine them.

The use of insecticide sprays can also cause poisoning in beehives that are located near crops. This must be avoided. To provide ventilation and more space, you can block the entrance to the hive. The restriction should be kept as brief as possible.

If the time is longer than one day, make sure that you have a container with water and a feeder with syrup. This is used to move colonies from one area to another. It is important to ensure that the area where the colony is moving is not near an area that will be sprayed. This is necessary to decrease bee loss.

Queens can only be raised from the best colonies that have good atmosphere and low swarming activity. This will ensure that good strains of bees are kept in the apiary. The queen should be used to re-queen and increase the number of colonies in the apiary.

You can keep some of the queens from the nuclei to use as emergency replacements for weak, missing or fallen queens. The hive might be too busy trying to build new queen cells, or raising new queens, and this could lead to a daily drop in population of around 1000 bees.

It takes bees about 15 days to find a queen, and 7-10 days to mat and lay eggs. It takes 21 days for an egg to mature into an adult bee, and 21 to turn into a forager. As you can see, the entire process takes a while. The colony's productivity will suffer as a result.

Your efforts in managing the colony from January to August will make a big difference in helping you harvest high quality honey. Light sugar syrup can be fed to bees along with supplementary protein. You should ensure that they are fed a balanced diet. Make sure they are protected from the cold and wind. During nectar flow, don't forget to stop feeding bees.

Chapter 6: The Different Members of The Colony

They are usually female but do not normally lay eggs. If they do, it could indicate problems within the colony such as a poor queen. An average worker only lives for six weeks.

There are two types of worker bees.

They collect honey and pollen. They go out every day to find flowers to collect. Once it has found a good supply, the bee uses a pheromone (a scent that guides other bees) to find the same area. One species of plant is all that a bee can collect in one day. A bee will only collect one species of plant per day if it starts its day by collecting from sunflowers. Because they don't waste pollen, bees can take it to other species.

Nurse Bees

These workers are those who work in the hive to care for the queen, feed the young, build new combs, and keep the hive clean. As nurse bees must be fed by foragers, too many could lead to a shortage of food. This is done by foragers who secrete a pheromone.

Drones

They are much larger than workers and have large heads. Although they do not work, they eat three times as much food as workers. Too many drones could lead to a decrease in food supply. They can be found only in the late spring and summer. As the temperatures drop, they are eventually forced from the hive and starve.

Queen bee (middle)

The Queen

Because she is larger than other bees and has a larger abdomen, it's easy to identify her, especially during egg-laying season. She does not reach the abdomen like workers or drones. Instead, her wings stop at about two-thirds of their length.

Queens lay eggs from spring through fall but few eggs in winter. A queen may lay up to 250,000 eggs per season. A queen can live up to five years but most live only two or three years. The queen's most important function is to produce pheromones, which are scents that help keep the colony's social structure together. Queen substance is the most important.

The life of a colony is dependent on pheromones. All the bee casts (workers and drones, queens), produce a variety of them to regulate many of their activities.

The Life Cycle of a Bee

As the queen gets older, her production of queen substances begins to decrease. This signal is sent to the colony by workers who create a special cell on the comb for the purpose of raising a queen. The larva that will be the next queen is given royal jelly, a special food. Royal jelly is also given to workers, but it is very small. The future queen, however, receives a lot.

The queen releases a sex hormone to attract drones when she hatches. She can mate up to 15 drones, who then die immediately. The queen then returns to her hive to start egg-laying. The sperm she produces from mating is kept in her body. It can be used to fertilize eggs of queens and workers, but not eggs that will become drones.

Reproduction of Bees

The queen lays the eggs in wax cells, which are the combs that worker bees build.

Each cell is given one egg. The egg hatches within three days and becomes a tiny grub.

The workers feed this grub with a mixture honey-pollen. The royal jelly is given to queen grubs, but the workers only get a small amount. The workers cover the cells with wax, and the grub becomes a. This is the stage where the adult bee grows. Depending on the species of bee, it takes between 12 and 14 days for the adult to emerge from the cell.

Swarming

Although swarming is an essential part of natural reproduction and the spread of colonies it can also be a problem for beekeepers. Problem number one is the fact that honey harvesting will be difficult after swarming.

Swarming refers to the movement of many workers with a queen or drone out of the hive in the hopes of setting up a new colony. Swarming behavior can be caused by many factors, but these are the most important:

The brood area is the part of the hive where the young bees are raised. There has been an increase in overcrowding. This is often due to overpopulation and lack of space.

Low production of the queen substance or low distribution through the hive due to overcrowding. Both cases stimulate workers to create more queen cells and raise new queen bees. These cells are a sign that your hive may be about to swarm.

Weather can also play a role. If there is a prolonged spell of bad weather, colonies that are strong and rapidly growing will be ready to swarm.

A failing queen, inadequate ventilation, a unbalanced age structure, and the fact certain varieties of bees are more susceptible to swarming than other types of bees are minor factors.

The colony will also be able to raise additional queens and reduce its foraging activity, as well as more drones. The queen will receive less food and her stomach will shrink, which will cause her to lose half of her weight.

The old queen will depart the hive in May or June during a calm, sunny afternoon. She will take with her at most half of the workers and some drones. Before leaving, the workers will have consumed large quantities of honey. This is the. The swarm will gather near the hive, which can be found on a tree, post or building.

Scout bees will search for a new location to start a colony, perhaps in a hollow or under a tree. The swarm will then move to the new location. They might spend a few days in the group before moving to the new location. This could also happen with some queens who are newly-raised leaving behind a smaller group.

Swarm Control

You want to avoid swarming because of its impact on honey production. It is essential to give enough space to allow the colony to grow. Most hives can add additional sections. If you see fruit trees and dandelions blooming in your area, it is likely that your time has come for an extension to your hive.

It is important to look for and remove queen cells in April and May. However, it won't stop swarming.

Splitting the colony and creating a new hive is the best way to prevent swarming. Although this is a complicated procedure that you won't need for a while with your new hive, you should still be ready to perform it if necessary.

Chapter 7: Your first hive

It is not difficult to set up a beehive your first time. It is not difficult if you know some basics. It is very similar to other subjects; you just need to know a few words and terms, and the utility of various items.

Before you even bring bees into your home, make sure you have chosen the perfect spot for your apiary. Your bees should never be placed in an empty space in your yard. There are many things you can do that will benefit your bees. You want to protect your hive from the elements, such as temperature, wind, sunlight, and sun.

Important things to remember when setting up a honeybeehive

Honey bees forage in a 3-mile radius from their hive. However, they will travel further if necessary and farther if not. Make sure that there is plenty of food nearby.

It is impossible to predict their flight patterns. You should make sure your neighbors are okay with your new hobby.

All local regulations and administrative requirements for producing must be met by the land where the hives will be located. Both the crops and the land must be free from synthetic fertilizers, pesticides and herbicides. Apiaries should only contain as many bee colonies as are needed to support the nectar- and pollen supply. You should also ensure that nearby farms do not use pesticides. (In particular, in particulate and dust form).

Face the hives towards the south/southeast if you can. This will give them more sunlight in the morning and protect them from the cold winds of winter.

Set them up so that some type of tree or brush growth is behind them. This will protect them against wind and severe weather. It also acts as a windbreaker. The image above shows an open area to the left. It would be wonderful if there was another tree. Or maybe a large bush.

The hives will thrive in dappled or irregular sunlight. They should be kept out of direct sunlight whenever possible, and they shouldn't be completely shaded. It is great to find a spot with dappled sunlight.

The hive should be level and slightly tipped so that rain does not get into the hive through the opening. The ground should also be level. It is not a good idea to put the hive on top of a hill, where it can be exposed to strong winter winds, or at the bottom of depressions where heat and humidity can accumulate.

The tricky part of setting up a beehive is after you have determined the ideal location. This guideline will help you. Remember that the ground conditions may differ from what you see here. Make sure to adjust accordingly.

To place the hive, make sure you have a solid and sturdy base. You must ensure that the base is sturdy enough to support the hive and doesn't wobble. It would be quite a disaster to tip your hive over.

Next, lay the bottom board in your hive box.

Place one Deep Super onto the bottom board. This is the brood chamber.

This deep can be erected with 10 frames.

With 10 frames, repeat the process for setting up another deep. This is the upper deep, or the food chamber.

These two chambers are your main beehive.

Your hive is now complete.

Chapter 8: Factors that Affect Honey Viscosity

Uncapped honey. Make sure the honey is well ripped

Some apiculturists don't wait long enough to harvest honey. This is the main reason honey isn't flowing. A good apiarist must be patient and allow the bees time to finish the honey comb before harvesting. This will prevent poor quality and fermented honey production. Different honey types: There are many types of honey, each one resulting from different flower types that bees eat. Honey tastes different. You can identify different types of honey by taking out your honey supers from a particular flower that stops producing nectar. Then, place new supers before the bees move to a different source.

The presence of eggs, larvae, and pupae within the honey wax can reduce its quality by creating bad taste, watering problems, and causing an abnormal smell.

Premature bee wax harvesting makes the honey watering.

Rainfall can cause the inability to obtain nectar or pollen.

Honey sucking can be caused by excessive bee wax accumulation in the hive. This also leads to heavy accumulation of eggs, larvae and eggs in the hive.

TEST FOR ORIGINAL HONOY

If the matchstick ignites, you can collect small amounts of honey. This indicates that the honey has been pure and original. If there is no flame, it means that the honey was diluted with liquid substances. Original honey can also contain white, bubbling particles such as in the case with palm win.

APICULTURE\BEEKEEPING-A PROFITABLE BUSINESS

Beekeeping is a lucrative business because the price for pure honey is extremely high everywhere. A super hive will produce approximately ten Litres (about 15 to 20 thousand naira) of honey each harvesting period. While some people view beekeeping as tedious and draining, others see it as time-consuming and dirty. However, it is one of the most lucrative and easy businesses a man can start. This is also called self-employment. The original honey also contains white bubbling particles, which are often found in palm win.

Once a farmer constructs the hive, supplies dunnages, and finds a dry, ventilated place to put the box up, he is done with beekeeping. Next, you need to check on the bee hive regularly to see if they are ready for harvesting. Fire outbreak is the biggest problem in beekeeping.

Apiculture can be a business that someone is interested in. However, it will allow him to have more time for other ventures. The farmer only needs to visit the farm once a year. Beekeeping is different from other farm activities such as crop and animal production. It requires regular inspections and proper care to ensure a better yield. Beekeeping is easy as long as the nectar and pollen are available, and that the environment of the hive does not catch fire. You will get plenty of honey.

Beekeeping is one the most straightforward agricultural activities, and it doesn't require a lot of money to start. The business can be started with a small amount of money, around twenty thousand naira. It is also easy to start. The farmer (apiculturist), must construct the hives and provide the smoker, dunnages, protective wears, and all other harvesting equipment necessary for efficient and effective harvesting and processing honey.

Chapter 9: Exploring the Different Species

Western Honey Bee

Apis mellifera (or the Western Honey Bee) is the most widespread species in the world and the most "domesticated by man to cultivate honey.

Honey bees have a variety of characteristics, including temperament, disease resistance, and productivity. A colony of honeybees also depends on its environment. Honey crops are dependent on the availability and quality of nearby plants.

The characteristics of a particular colony are affected by the genetic makeup. This allows beekeepers the ability to choose strains based on their priority: pollination, honey production or honey crop.

You can think about the characteristics that distinguish certain groups of bees as a beekeeper. Be aware that stocks can be very diverse. There will always be exceptions to the rule.

Beekeepers will be better able understand the various types of bees available by simplifying. There is no "best" bee strain. It all depends on the individual beekeeper's preferences. This could be different from another beekeeper's first choice.

These subspecies are the most sought-after and are among the most common honeybee stock in the U.S.

European Dark Bee

Apis mellifera, the European Dark Bee. Sometimes they are called the German Black Bee, or German Dark Bee. This subspecies is European and originated in Europe. It can be found from Western Russia to Northern Europe and down to the Iberian Peninsula.

The European Dark Bee was domesticated and introduced to North America by early European settlers in 17th Century.

These bees are quite large for honeybees. Their bodies are large and have thicker, longer hair. This allows them to stay warm in colder climates.

The European Dark Bee's color is dark. They appear blackish to the naked eye, but they are actually a deep dark brown.

This subspecies was once the dominant species in the feral bee population of the United States. New diseases have almost eliminated the majority of wild bee colonies.

Positives

This is a hardy strain that can withstand cold winters in all climates.

Low Swarming - This makes it easier for beekeepers to manage them.

High tendency to collect pollen.

Queen and worker bees have a longer lifespan. Queens are not productive and do not require replacement. They are able to rule the colony throughout their lives and do not require replacement.

Negatives

Stock of the German bees is currently not readily available due to their drastic decline worldwide.

New diseases, such as American or European foulbrood, can make you vulnerable.

German bees, especially hybrids, can sometimes be overly defensive. It is therefore more difficult for beekeepers to manage this bee stock.

Italian Bee

Apis milliner Linguistic, also known as the Italian Honey Bee or Apis mellifera linguistica, is a native of Italy and the surrounding Mediterranean. Sometimes it is called the Ligurian Bee. They were introduced to the United States in 1859 and quickly became the most beloved stock in the country. This is still the case today.

This honey bee subspecies is one of the most widespread and adaptable. It hasn't thrived in tropical areas and is less able to withstand cold winters.

There are yellow and brown bands on the abdomen. There are three colors available for Italian bees: leather, bright yellow (golden), and very pale yellow.

Their bodies are smaller and their hair is longer than the darker honeybee races.

Positives

A gentle and docile disposition

Prolific breeding

Comparable to Western countries, there is a lower tendency to swarm.

Honeybee races

High standards of cleanliness and housekeeping are essential. This is thought to be a contributing factor in disease resistance.

The honey is covered with white "cappings".

Amazing foragers

Most people are willing to take part in supers.

Negatives

The Italian bees can eat excess honey from the hive due to their brood rearing. This is particularly true if the bees are unable to forage enough.

They have a tendency to attack other hives and steal the honey. This could lead to spread of transmissible diseases among hives.

Sensitivity to diseases

Carniolan BeeApis melifera carnica, the Carniolan (or Carnies as short), or Grey Honey Bee or Slovenian Bee.

This subspecies is native to central Europe and is smaller in size than other European bees. It is extremely gentle and colonies grow back quickly as the weather warms in spring.

This subspecies is second in popularity among beekeepers, and highly respected by U.S. beekeepers.

Positives

Carnies are calm, gentle, and non-aggressive. They can be kept in areas with lots of people. They can be used with minimal smoke and protective clothing, according to beekeepers.

The availability of nectar is what these bees use to adjust the worker population. The population rises rapidly and extensively when nectar is readily available in Spring. In contrast, brood production and brood rearing are affected by colder weather, when nectar is less plentiful.

They have high worker numbers when there is high nectar availability. These birds are able to store large amounts of honey and pollen over these times.

The Carniolan bees can also resist parasites and diseases that could cause the destruction of colonies of other species.

This makes it less likely that other colonies will be raided and their honey taken. This helps to reduce disease transmission between colonies.

Workers bees live an average of 12% longer than other species.

Excellent builders of wax combs. This is useful for cosmetics, candles, soaps and other products.

Negatives

Carniolan bees are known for their strong instinct to swarm when they see too many people. This can make it difficult for them to manage. The beekeeper may end up with a poor honey crop. To prevent the loss or swarms, the beekeeper must be extremely vigilant.

They tend to die in very hot summer temperatures.

The availability of pollen is a key factor in the strength of a brood nest.

You are the beekeeper and it is difficult to find the dark queen.

Caucasian Bee

Apis mellifera Caucasica; The Caucasian Honey Bee.

They are found in the Central Caucasus' high valleys. The subspecies are primarily found in Georgia, but can also be found in Armenia, Eastern Turkey, and Azerbaijan.

This stock was once very popular in the U.S. but has declined in popularity over the past few decades.

The Caucasian Honey Bee can be large and mildly colored. Due to the thick hair covering, it can appear gray. Some people have brown spots.

Its most distinguishing characteristic is its extremely long tongue, which is the longest of all bee breeds. This allows bees to find nectar from flowers other bee stock may not be able reach.

Positives

Extremely docile and mild-natured

Strong colonies are formed by enthusiastic brood production. The availability of nectar affects the colony's ability to regulate brood production. It is therefore highly productive and produces high honey yields.

It is best to find areas with high nectar flows in the middle of summer. Because colonies reach their full strength in the middle of summer, this is why it is so good.

It uses a minimal number of combs for honey storage. The comb is fully filled before the new one is started. This saves time harvesting honey.

It can withstand rainy, cold and unpredictable weather.

Negatives

They are unable to produce large honey crops due to their gradual spring buildup. They are not suitable for areas with high nectar flows in the spring or early summer.

Hive management can become more difficult if propolis is used in a fervent manner. This basically means that frames and boxes can become very stuck together. Propolis, also known as bee glue, is a sticky resin substance that seals cracks and joints in bee structures.

Tendency to steal from other hives and drift.

Buckfast Bee

This is a "manufactured" bee breed that was created from the cross of several strains of bees. Karl Kehrle, also known as "Brother Adam", created the Buckfast. He was a monk in Buckfast Abbey, Devon, England where he was responsible beekeeping. There are still bees bred there today.

In Great Britain in the 1920's, tracheal mites decimated bee colonies and killed thousands. Kehrle was given the task of creating a bee population that could withstand this deadly disease.

There were 16 colonies left in Buckfast Abbey, which was only one of the remaining colonies. These colonies were made up of Italian (and Italian-derived) hybrids of German Black Bee and Italian Black Bee. Kehrle also imported additional queens from Italy. This work resulted in the Buckfast Bee.

Positives

Calm temperament and docile disposition, with low instinct to sting.

Good honey crop production

Strong resistance to diseases

Foragers who are hardworking and passionate.

Very low tendency to swarm.

This helps to lower the risk of getting sick.

Capability to thrive in both cold and wet environments

Negatives

Restricted brood production during cold seasons

If a colony isn't re-queened, the Buckfast Bee may become overprotective in a second-generation colony.

The spring population growth is slow. They are therefore unable to fully take advantage of the early nectar flows.

The Russian Bee

This stock is among the earliest to be imported to America. It was originally from Russia's Primorsky Krai.

Infestations of parasitic mites caused a dramatic decline in the bee population in the United States during the 1990s. Scientists discovered that Russian bees are resistant to many parasitic mites. The USDA's Honeybee Breeding, Genetics and Physiology Laboratory used them to strengthen and improve existing stock.

The USDA conducted tests to determine if the stock was resistant to varroa. This project's quarantine phase was completed in 2000. This strain of bees has been commercially available since 2000.

A Russian bee's genetic makeup is Caucasian with some Carniolan and Italian lineage. They share some similarities, but they have their own unique traits.

Russian bees are dark in color. The Russian bee's abdomen is black and the hair is gray.

Russian bees will only rear brood during periods of pollen and nectar flows. The environment can have an impact on brood rearing and the population of colony members.

This stock exhibits some unusual behaviors. Queen cells are present in the colonies most of their lives. If the queen is not available, the queens can quickly emerge and take over egg-laying duties. They act as an emergency backup queen in case of trouble. Other stocks, on the other hand, rear queens during times of queen replacement or swarming.

Positives

Gentle disposition with low inclination towards sting.

They are resistant to many diseases, including varroa. Their excellent housekeeping may partly explain this.

Good honey production.

Negatives

The tendency to swarm, although this can be managed.

The majority of Russian queen bees on sale are hybrid daughters from a breeder queen. This means that the Russian queen has openly mated any drone with which she may have come from. The result is a genetic hybrid. Recent research has shown that hybrids are only partially resistant. Research at North Carolina State University found that even partial resistance to mites is significant and important, particularly when compared to other stocks like the Italian.

Only by isolating the breeding ground so that the drone stock can be controlled can pure Russian queens be made sure. This was what happened at USDA's laboratory.

The Giant Honey Bee

Apis dorsata, the Giant Honey Bee; is found in south and southeast Asia. Although the species can be aggressive when provoked it is a prolific honey-producer. The honey is harvested from large, exposed combs by local indigenous peoples. However, the species has not been officially domesticated.

Africanized Honey Bee

Apis mellifera cutellata, the Africanized Honey Bee is also known as 'killer honey bee'. This species is a central African native and has a reputation for being aggressive. However, it is not more aggressive than any other species. Its reputation was tarnished as a result of 1956 breeding experiments.

Brazilian geneticists were trying to breed a new hybrid with the hope of producing more honey. These super-defensive African bees were crossed to various European bees that had a gentler temperament.

Unfortunately, some African queens bees managed to escape into the Brazilian jungle. These queens crossed with jungle bees to create the 'Africanized Honey Bees. These colonies had unusual defensive tendencies and kept guard bees between 20-30 metres from the hive.

These bees were sent out to protect these colonies. They were numerous and persistent. They traveled long distances to pursue the intruder.

The hybrids were also able to take over European bee colonies which resulted in rapid hybridization and reproduction.

These hybrid bees are known as the "Africanized" honey bee. They have spread through Central America and Mexico to now reach the southern United States.

The Africanized Honey Bee looks very similar to the friendlier European Honey Bees in terms of appearance. You will need to perform a DNA test on the samples or examine them under a microscope in order to detect any differences.

It is important to note that they don't sting more severely and do not die after being stung.

Their temperament is what makes them different. They are very protective of their hives, and they react quickly to any disturbance that they perceive as a threat. They can chase an intruder for a long distance and remain hyper-defensive for days after an incident.

Chapter 10: The First Season

A swarm of birds is a good idea.

Catching a swarm is extremely easy. It is prudent to dress appropriately when you are trying to find a crowd. Even if they've only been around for a few days, honey bees can become very aggressive. If you spot a swarm, it is possible to spray fresh water with a brush or a shower container to stop them from wandering around. The swarm is kept under a little hive with no outlines. A crate or case can be used. A honey bee cloak or a swarm cloak may be used for this purpose. The honey bees will fall into the hive by blowing on the branch from which they are hanging. Spread the colony using a lightweight material and place it in the shade. If a firm blow does not have an impact, you may also drive the honeybees into their hive using a honeybee brush or smoke. The queen will be the only honeybee in the colony. You should take a small section of the edges out of a hive and then shake the honeybees into the big colony. Once the colony has been closed, the casings must be reattached. The colony's flight passage can remain open. Honey bees that are still in the colony may be tapped onto the flight load-up. You should strengthen the honey bees the next day. This will be something you'll continue to think about.

Baiting a swarm

A little hive just occupied by honeybees can be taken. This will make it a flexible apiary. Fill it with top bars or casings. Two edges should have brushes. The other edges should contain establishment sheets or pieces of an old toothbrush. To provide some protection from the wind, place the hive on a roof or tree. The honey bees will begin to organize themselves around the situation of the colony once the swarm has overtaken it. The swarm should place the colony in the best spot possible as soon as it has taken control. The honey bees may only have been involved in the colony for a few days. In this case, they will arrange themselves to the colony. It is possible to move the colony over a short distance, but it may not be possible to transfer it back to its original location. If you happen to have any old colonies, a loaded load with a spreader and a baseboard is sufficient. Keep one around your home to attract a swarm. The swarming season is usually in spring, or towards the beginning of the dry season. It is not worth trying to control a large number of bees. It is important to know that honey bees are not always available in all countries.

Collecting wild nests

You may find old swarms under trees in a woods, or elsewhere. These are bees that have just settled and have made a few brushes. On these brushes you will find brood and honey. Although the chances of this state remaining in your beehive after catch is slim, it can still be seen by trying to move the whole colony into a hive. To reach the brush, you will need to open the tree. To get the honey bees out of the bush, use a lot of smoke. Then, you can remove the brush with a sharp knife. Trim the brush to fit the casing. You will need to cut through the cells where the cables are to be run. This will allow the cable to pass through the brush. You can also attach a string to the edge and bush as additional protection. These brushes are attached to the sides and then hung in the bee hive. Finally, the honeybees will be able to crash into the colony or are cleared. These honey bees are just now able to locate themselves. The colony should be moved around 5-6 km from where they were discovered.

A larger hive

When every brush in the colony has been loaded with brood, it is time to make more space. It is possible to make an augmentation to your hive if the colony is becoming too small. If you're using removable brushes, move 50 cm the small hive to the side and place the large hive there. Light some smoke on the small colony's casings. After a while, sit down and then release the casings using the hive tool. The principal outline should be held by the sidewall, with the handles close by. Next, gently place it in the large hive. You should move the different casings according to their request. The goal is for the brood home to be the same shape as the original. You should look for extra edges along the sides of the moving frame.

Administration

A decent record should be kept of the condition of your settlement, especially if you have several colonies. After every investigation, you should keep a record of the date, the proximity of brood brush, the nourishment supply. This includes drone and swarm cells as well as any moves made. You should also note the honey yield or nonappearance, as well as any other relevant points, such forcefulness, etc. The hive card can be connected to the underside or front of your hive. You can also use the card framework to write all the details in an activity book or a leaf document that you bring home. If you don't have gloves, it can be difficult to write on the hive cards. It is also more simple to take a few notes on paper and then to make detailed notes at home. It is important to count the hives for authoritative purposes. You will find that all the information you have gathered is very useful when there are more colonies. Then, it's time to start choosing the best.

After seven days, inspect the hive. You should not disturb the honey bees too often, but as a fledgling you still have a lot of work to do to ensure their survival. The colonies should be inspected during daylight, when the weather is warm, but not when it is raining. Open the hive carefully and exhale a bit. After a while, remove the spread fabric from the casings. To release the casings, use the hive device. Take a few puffs of smoke from time to time. After balancing the leading edge of one handle against the beehive, take out different housings and inspect them individually. As much as possible, ensure that the benefits are spread by the wet-spread fabrics.

Pay attention to the following: Are there eggs? Hatchlings? Top worker brood? Drone brood. Is the queen present at all? Is there enough food? Is there any evidence of wax moth hatchlings in the area? Are the honey bees or brood still solid? These discoveries will be noted on the hive cards. The queen should not fall beyond the hive. You must keep the edges of the hive in check. To prevent illness spreading, especially with regard to American bees,

Foulbrood is used to disinfect equipment before inspecting different colonies. Between apiary visits, gloves should be washed before going to another. The exchange of spores between the colonies can also be prevented by heating the instruments of the hive in a smoker.

A support area is provided to stimulate improvement during periods of inadequacy. The brood is strengthened by the regular continuing with small amounts of sugar (or debilitated Honey). Although the sustenance provided by the appearance is vital for the continued existence of the bumblebees, it doesn't allow them to develop into a more distinctive species. Without sugar support, a region where you have just removed honey cannot interface with a need period. To make a sugar course, heat equal amounts of sugar (extraordinary value valuable stone sugar) and warm water. Stir until the sugar separates (don't bubble). Use lighter sugars to avoid bumblebee runs. A feeder is used to support the bumblebees. A large jam container or small plastic bowl can be used for this purpose. You should make a lot of holes at the top, each about 1 mm. To pound the openings in the jam holder's metal top, use a nail. You should make an opening that is at least one degree smaller than the feeder on the inside of the hive. Over the chance in the inward spread, locate the feeder with its punctured Top upside down. You can then place the outer front of your hive on top, and a honey super or brood chamber will be placed over it. The feeder can be placed in the place of several unutilized edges in the colony. You can also place the feeder over a small wooden plate that you place at the flight entrance. You must ensure that the bumblebees don't land on the sugar game plan, taking all factors into consideration. This system is empowering in that the hive should not be opened by any means.

You must ensure that there are no openings where honey bees and wasps or ants can get in to steal the sugar. By making the entrance to the flightway smaller, you can prevent burglary.

Don't make more sugar arrangements than honey bees can eat in a few days. Honey bees can be hurt by an old sugar arrangement.

If the honey bees don't take up sugar again within a few hours, it is time to stop bolstering. You should close the inward spread at the time you have evacuated the feeder.

If you are able to sell honey at a reasonable price, it is worth sustaining the bees in the midst a reduced honey stream. The subsequent growth of the colony, and higher honey yields will help you more than offset the cost of sugar.

You can't give food to the state if you want to avoid burglarizing.

Sugar can be nourishing and will help to stabilize your state during periods of low pollen. You may find no pollen in the hive. The absence of powder means that the brood is not receiving enough nutrition. This will result in fewer honey bees. You can offer the honey bees soybean flour as a substitute for pollen. You can mix the substitute with some sugar to create a protein-rich cake. This can be encouraged by placing it on top of the bars. It is best to not make too many cakes as they are usually short-lived.

The management of the colony's growth

The brood home spreads continuously over many brushes. Establishment sheets and strips are the basis of brushes. Brood, pollen, and honey are added to the cells at the edges. You should provide small amounts of sugar arrangement to reduce building costs and vitality. Once the honey bees have filled all of their casings with brood and nourishment, it is time to add more room. The type of colony you have will determine the best way to give extra space. If the honey zone is located in a long hive or top bar hive, where it is adjacent to the brood room, the honey area will be separated from the brood compartment by a piece of hardboard or similar material. The honey territory's flight path will have been closed.

It is vital to keep increasing the size of the colony (number and type of chambers) by adding more. Honey bees should have the ability to reach all edges. The entire brush surface can be protected from gatecrashers (for instance the wax moth), and maintained at the right temperature.

This standard is essential for tropical honeybee races such as the Apis cerana and African mellifera in Asia. Every brush should be covered with thick honey bee drapery. A beekeeper should also have the ability to reduce the number of bushes if the colony becomes smaller. The wax moth could genuinely attack the home and steal the colony.

There are many reasons why you might fall off the track.

Attack by ants.

Looting is another activity of honey bees.

Separation was caused by a too broad or tight brush.

Frequent and severe aggravation (for instance by the beekeeper).

Place a queen excluder between your family and honey parts to prevent brood from forming in the honey chamber. It isn't important. You can encourage workers to quickly develop honey super fast by placing a few casings with topped families (ensure that the queen is not on the edges) in honey super. You can replace the brood home casings with similar advantages using brush or a sheet. If they are similar in size, casings should be traded between the honey chamber and the brood chamber. The honey stream and climate are key factors in the rate of improvement. A number of honey bee scrounge plant are found in the area, so if the environment is favorable, the young state will flourish. In the event that a colony is not growing well, you should check for inadequate nutrition and water quality. You also need to check for honeybee infections.

Honey flows

Honey stream is when multiple plants that give honey bloom at once. A honey stream's wellspring can be one type of plant, such as unadulterated Eucalyptus honey, or many types. A few elements in the plant world trigger blooming. This is mainly temperature and the length of sunlight in temperate zones. The tropics are where the temperature and length of the day don't change too much, so blossoming is often controlled by the dry spell and downpour. Most plants in the moist tropics bloom after a few weeks of dry weather. There are some exceptions. For example, citrus varieties that bloom after torrential downpours have started. Arid areas thrive in the middle of the stormy season.

The beekeeper should note which plants the honey bees fly to. A novice may think it helpful to note, for the first few years, when honey bee rummage plant blossoms. You will be able to get a good overview of the honey year and make adjustments for the honey stream.

Honey gathering can start three to a month after the start of a decent honey stream. The honey can then be concentrated and the fixed brushes removed. If you have diagrammed a hive, you can leave the entire housings until the end of the honey stream. In any case, you should give the honey super fast so that the honey bees have enough room to store it. The brood chamber should not be removed from the baby. The baby should be kept in the brood chamber for honey bees to use during periods of shortage.

The rest periods

The stream will check the amount of honey left to ensure that the honey bees are getting enough nutrients. If you haven't yet removed any baby bees from the brushes, the honey bees will still receive sufficient nutrition during the shortfall time. If you have not yet collected baby or the colony is too weak to provide adequate honey stockpile, you can continue to give little sugar arrangements. You should also check that honey bees have not been disturbed by termites, ants, or wax moths during this time. Close the flight entrance using nails or coarse work to keep mice and other reptiles out. The honey bees will usually get out. As honey bees can't protect them against wax moths, all brushes must be removed. The wax moth can form into a plague by being near vacant brushes, and eventually the colony will leave the hive.

Frames with comb

It is difficult to keep void brushes in good condition when you live in the tropics. It is best to store edges with brushes in a secure and ventilated area. You can cover the pile of void brushes with mosquito netting so that at least 20 cm is separated between the cloths and the bushes. Keep new honeycombs with a light shade, as flat brushes can be destroyed by parasites. Cashings should not be stuffed tightly on racks or hung on wires. Keep outlines with brushes out of reach of wax moths.

Chapter 11: A Look At Benefits And Challenges

The current pressure on bees is immense. There has been a rapid and dramatic loss of colonies in many countries around the globe, for reasons that we don't yet understand. It is a very frightening possibility. We risk losing one of our most important pollinators and causing massive damage to the human community. Earthwatch and the Royal Geographic Society declared the bee the most important animal on the planet in 2019. In recent years, ninety percent of all bee species have disappeared. The most striking aspect of this story is the apparent lack of concern among most people. We don't know the implications of the bee disappearing, so we aren't able to fully understand them. This is why much of the world seems to ignore this fact. According to a 1998 estimate, bees pollinate 200 billion dollars worth crops each year. Between 2007-2013, 10 million colonies were lost. This is twice what we would expect under normal circumstances and represents the worst loss of beekeeping history. It's easy to see why losses at this rate could quickly become unsustainable.

There are many theories about the cause of the sudden decline in honeybee numbers. However, we can debate them. It is clear that there is more to be done to save them. Although scientists, beekeepers with professional credentials, and agricultural officers are trying to solve this problem, it is surprising that many of the efforts have been made by small-scale beekeepers. Since the widespread publicization of the colony losses, backyard beekeeping has experienced a huge increase in numbers. This may be a good thing, as these home-based operations could become an important reserve if the current downward trend continues. Instead of large numbers of bees being kept close together by beekeeping companies of industrial size, thousands of smaller hives are spread more widely. This may help to save bees from dying from transmittable diseases.

Colony collapse disorder usually results in worker bees disappearing. Keepers open their hives and find the queen, as well as plenty of food, in the midst of the rest of the colony. The problem is that bees can be tiny, so millions could die. Nobody would know where or why. Workers bees travel great distances to find food. Therefore, even large numbers of insects scattered across a wide geographic area would not be easily spotted.

CCD can be caused by many theories. Two possible causes of CCD are Varroa mites and Acarapis mites, while another is immunodeficiency. Although global weather change may be a factor in some of these problems, it is unlikely that it will cause such a drastic loss of colonies. Bees are primarily used for pollination in the United States, and other countries as well. They are often transported long distances because of this. They might be transported to one location to service a particular flowering crop and then again to another area to do the same for a subsequent flowering crop. It may happen several times per year. It can be stressful to move and relocate at each destination. This could also have been a portal for a new disease or other health issue. This possibility is possible, but we know that colonies that do not have to be moved are being lost. The possibility that transported bees have contracted a disease and then transmit it to other colonies is possible. However, the exact cause of the disease remains to be determined.

The pesticide family of neonicotinoids is probably the one that has attracted the most attention. The Shell Corporation first discovered neonicotinoids. However, many other companies produce neonicotinoids-based insecticides. It seized 24 percent of the global pesticide market within a few years of being first made available to the public in 2008. Sales grew from $155 million to $957 million annually between 1990 and 2008. It was used in the USA as a seed coating for 95 percent of canola and corn crops, and almost all of the vegetable and fruit crops. The short version is that neonics were almost everywhere in the world of agricultural crop production at the same time as the dramatic drop in bee numbers.

This is a contentious issue in science, with both sides pointing fingers at each other. While the environmental lobby insists that they have identified the cause, the big pharmaceutical companies claim that there is not enough scientific evidence to back their claims. The majority of government bodies in the EU voted to ban neonics in 2018. The US didn't feel obliged not to be cautious. Although the Obama administration did prohibit use of farms in National Wildlife Refuges, that ban was overturned by Trump's new administration.

Representatives John Congers, and Blumenauer presented the Save American Pollinators Act to Congress in 2013. It was then sent to the congressional committee, where it never left.

The argument is too technical for those of us not trained in science to make an informed decision. Both sides have a lot of lobbying power and both sides are able to export jobs and money. Although it would appear that the European decision of following a cautious approach until more evidence is available is wise, that opinion is not shared in the United States.

The above information may be confusing for a potential beekeeper trying to decide whether or not they should invest their time and money in a few hives. Your efforts may be vital now more than ever before. It could be possible that CCD is not affecting you if you live far from neonics-using areas, such as in a city or town. If this were to be the case, and it was repeated in other areas, it would indicate that pesticides could be the cause of the problem.

No matter how much chemical use has been made, beekeeping will continue to present a challenge. Beekeeping must be free of disease and mites, and protected from rodents and other pests. All of these issues need to be taken into consideration, but the benefits must be weighed against them. These threats are well-publicized and visible. In a world that is dominated by the negative, the benefits are much less tangible and will always be overlooked.

When you start learning about beekeeping, there is a steep learning curve. It can seem daunting. You should keep going. It is a natural process and you will soon find things flow naturally as you engage with them. You don't have to learn all there is, but it is better to take your time and grow than to try to memorize everything. For millennia people have studied bees. Today, they are one of the most widely studied insects on the planet. We seem to be learning from them.

While most subjects can be tedious, learning beekeeping can be so enjoyable that you will absorb the information almost like osmosis. To regain touch with nature, you need to learn just one thing. Being a beekeeper is about learning about the pests that can threaten your hives and the different flowering plant species that support them. Your beekeeping skills will improve, as will your appreciation for the environment they live in. The area will have more fruit trees and flowering plants, which will in turn provide habitat for other animals. I'm sure you will feel compelled to quit using pesticides and may also desire to avoid herbicides. You will soon discover that there are natural alternatives to these pesticides. And so, the learning circle grows.

If you don't like being alone, beekeeping allows you to connect with other people who are interested in the same things.

They will initially be your guides but, if you persevere with the project, you'll soon find yourself helping others. One word of caution: Beekeepers can become passionate about their topic and want to spread the word to anyone they meet.

It is not costly to keep bees. Even if you are only selling honey to your friends, any outlay can be quickly recouped. The majority of honey found in supermarkets has been altered in some way. You can be sure that your honey is pure and you know the origin of the flowers used to make it. This will ensure that there will be a steady market for any honey you don't eat.

Children are always fascinated by the contents of those boxes in your back yard. Even though you might think you are boring, your kids will be enthralled by the sight of the boxes in their backyard. You will be their most interesting adult, unless

Superman shows up. Although I don't have proof, I believe Superman is a half-decent beekeeper.

Children learn quickly to overcome their fear of being surrounded with hundreds of buzzing insects and become dedicated students. You will help to create the next generation of eco-savvy children who will hopefully make the planet a better place.

Let's recap briefly: Beekeeping is a way to help the environment and reduce the harm we do to it. You will learn a lot and produce better vegetables, fruits, and flowers. It can help you make friends, get along with your grandkids, and provide you with many useful or saleable products. You don't even need to be a masseur, and beekeeping is one of the most relaxing activities you can do.

There are many career options for those who have become addicted to beekeeping and want to make it more than a hobby. There are many options. One is to increase the number of bees you have and become an apiculturist full-time. But that is not the only choice. For those who are interested in the science and technology of the industry, there are many opportunities for research. If you prefer to be hands-on, then there are often opportunities for beekeepers who work on a seasonal basis at larger operations. Keepers may spend their summers in different parts of the world, working on bee farms in the US or New Zealand. A wide variety of satellite businesses exist, including teaching, speaking at events and writing about beekeeping. Even though it is not recognized by the US, there are individuals who can make therapeutic medicines using the numerous products each hive produces.

It is clear that I have painted an idyllic picture of beekeeping by focusing so much on its benefits. To be fair, I will also discuss some of the disadvantages. As good farmers have a relationship to their animals, so beekeepers should also have a relationship to the insects they care for. You'll know which hives are aggressive, which ones will swarm soon and which ones are not thriving as well. This knowledge will eventually become intuitive. It can be very depressing when something goes wrong. You will often come home from inspecting your beehives and vow that you will never keep them again. Sometimes the threats to bees seem overwhelming. The hive is vulnerable to attack by mammals, insects and viruses as well as birds, mites, mites, and fungi. Pesticides pose a lingering threat that we may never fully understand. As you read this book, you will discover more about non-chemical barriers. All beekeepers await the results of investigations into the chemical threat. We can only do our best to keep our bees safe from any man-made chemical threats.

Chapter 12: Benefits of Bees in an Orchard

Many people might be puzzled by the uneventful scenes among rows and rows upon rows of trees. They may also misunderstand the labor required to run an apple orchard. It seems that everything else is quiet, aside from the bustle of workers picking and harvesting apples in summer. They often mistakenly believe that operating an apple orchard is so simple that it would require little knowledge.

It is hard to believe that owning and managing a fruit orchard can be a difficult and tedious task. It is not enough to think twice about the maintenance of apple trees. The size of the orchard is another thing to consider. According to horticulturists, an apple orchard must be at least ten acres in size to make it profitable.

Imagine how many acres of trees are needed to make a profit. An orchard that is larger, more productive, and more profitable financially will have more trees, more land, and more labor. The cost of equipment, insecticides, and other tools necessary to ensure crop security, as well as the costs for employees, can be quite expensive. This is not to mention bees' role in orchard success.

New orchard owners often make the biggest mistake when their apple trees are blooming. All blossoms must be pollinated in order for apple trees to bear any fruit. Each tree. Many orchards die within the first few year of being established. This is because there are not enough bees to pollinate all the trees. Local beekeepers are a must-have for any orchard owner.

Many beekeepers will let hives and honeybees to growers. Many beekeepers come to the orchard and set up their hives amid the rows of apples trees. The wild bees benefit from the added bees. Pollination is an essential part of producing healthy and high-quality apples. Cross-pollinating apples was possible only with the help of bees. Cross-pollinating occurs when bees fly from pollen-producing (or donating!) trees to pollinate other trees. Cross-pollination must be achieved by planting apple trees in rows that are alternating, with crabapple trees used to spread them out.

The morning hours are when honey bees visit pollen-producing plants. This is why it is crucial that apple orchard operators and owners do not interrupt the early morning pollination process. Insufficient pollination can cause apple trees to develop deformities and early fruit drops. This can be detrimental to the tree's ability to produce fruit. Their flowering stage typically takes between 8-10 days after apple trees have been properly pollinated. However, cooler temperatures can extend this time.

Honey bees are one of the most important pollinators for orchards. Many orchards have on-site beekeepers and high pollination service providers to help them find quality colonies. Many growers want to encourage wild bee populations in their area by restricting the number of flowering areas.

This practice is cost-effective but requires many considerations in order to promote healthy bee populations for pollination. Cultivators must plant non-crop flowers in the areas near their orchards that have not been treated with pesticides and are not subject to large amounts of pesticide drift. Orchard growers need to be able to provide information on how to care for the native bee population. This includes where to store them and when to take them out.

The season.

American beekeepers are also interested in California's almond industry. Almond trees cannot self-pollinate so bees are required for a productive harvest. There is a huge demand for honeybees, and beekeepers are experiencing significant economic growth.

Many beekeeping companies in Florida have started to cancel or evade contracts with watermelon farms nearby to bring their services to California. Prices for beekeeping services have skyrocketed, as well as the prices almond orchard farmers are willing to pay for hives to provide pollination services. California is the only region of North America where almonds can be grown commercially. This is due to California's climate.

In order to pollinate almond trees, beekeepers need to place their beehives around orchards when they bloom. This is usually February.

Chapter 13: When and Where to Setup Your Bees

This chapter will help you choose the right site for your hive, and how to get started.

You don't need a lot space to set up a beehive. You can be a beekeeper no matter where you live, whether you're in a large city or on a farm.

What to Look for in a Site

Accessibility: The hive should be easy to reach so you can inspect it. If you want to set up your hives on the roof of your house, how easy will it be? How convenient would it be to set up your hives in a field? Remember that harvesting involves carrying honey frames and heavy loads.

*A water source: Just like us, bees also need water. It is used to cool the hive and to drink water, as well as to dilute honey, if needed. You should ensure that they have water access. If they are in dire need, you can place a large bowl with water on their table.

*Well-drained: Make certain that your chosen area is not likely to flood. The water will drain quickly. A hive can be very heavy and could sink if it becomes a quagmire in the rainy season.

*Sheltered bees need shelter from the elements. They prefer a hive with a steady temperature. The bees can become aggressive if the hive is too hot. If it gets very bad, they may die. They will stop producing as much if it is too cold. To keep the temperature constant, invest in some shade cloth if you're an urban beekeeper. You should also ensure that the hive is well ventilated and protected from any storms.

*Face the rising sun: The bees will work much more efficiently if they are allowed to face the rising sunlight. This increases the amount of time they can make honey.

*The site should be level. However, it is a good idea to lift the back a little to let water drain out.

Keep the entrance clear. Placing a layer or plastic of mulch directly in front will stop grass from growing and blocking the entrance.

*In the city, research your hive location before you get started. Is there a place you can set up your hive in an area that isn't too far away? Check to see if the hive can be placed on the roof of your home. In case of disputes, it is a good idea for you to obtain written permission.

How to Set up Your Bees

This will usually occur in the early spring when the flowers start to bloom. This will depend on where you live. You should have the hive ready before you go to get your bees.

This chapter teaches you:

*How to select a site that is suitable for your hives.

*The best time to put up is early spring.

The next chapter will discuss the equipment that every beekeeper requires and some useful extras.

Chapter 14: Ancillary Equipment

A smoker and a hive instrument are essential for working bees. A smoker is a metal firepot with attached bellows, grate and chimney. The size of the smoker is a matter of personal preference. The 4x7 inch scale is perhaps the most popular. If you intend to support the smoker between the legs of your colony members, you will need to purchase/use a smokestack with a heat shield. A hook is preferred by some beekeepers to allow the smoker to be hung over the open hive. This will make it easier to inspect the smoker and keep the smoker close at hand. To produce thick, cool smoke, coal must be at the top of the grate. Burlap, corn cobs and wood shavings are all suitable smoker materials. You can also make liquid smoke by mixing water with it and spraying it on the bees with a mister-type application. If theft is not possible, you can mist with sugar syrup and smoke.

Hive tool – The hivetool is a key tool for prying apart frames within a honey mega or brood chamber, as well as separating the hive body and scraping out wax and propolis. Although there are holsters that can be used to carry hive equipment in your hand, many beekeepers prefer to hold the tool in your palm. This allows you to open it and allow you to reach boxes and frames. It is important to wash the hive method of extracting honey, propolis, and wax. You can do this by either putting the tool in the ground or using a hot stove to heat it. All cleaning products can be used to stop the spread of bee diseases. A screwdriver and a putty knives are not good substitutes for a strong hive device. They can cause damage to the frame/hive bodies.

Protective clothing: You will wear a bee veil on your neck and face to protect yourself from bee stings. There are three types of veils: veils that open at the top and can be worn over a hat; veils that cover your entire head completely; and veils that form part of a bee suit. The best protection is provided by a wire or cloth veil, worn over a wide-brimmed, lightweight hat. Even though they are lightweight and easy to transport, veils without hats do not always fit as snugly on the head. The elastic band that wraps around your head when you are trying to get rid of bees acts backwards and causes the veil to fall across your hair and face as you lean over.

There are many options for beekeepers to choose from when it comes to covering their bees. Even the most expensive bee suits may not be the best or easiest to use. If you wash your clothes frequently and keep them clean, coveralls can be useful to reduce the chances of getting propolis on your clothes. Beekeepers will appreciate the use of shirt veils (long-sleeved shirts) or coveralls. For working bees, white or brown clothing is better. Although many colors are acceptable, bees do not like dark colors or soft fabrics. While windbreakers and coveralls made from ripstop nylon cloth work well for bee workers, they can become too heavy in summer. For beginners who don't like getting stung, rubber gloves or cloth are recommended. Some beekeepers are uncomfortable with gloves and will take several stings to make it easier to handle the honey. Form-fitting gloves, such as those for domestic or laboratory tasks, reduce honey and propolis stings.

Exposed wrists and dark socked ankles are more susceptible to bee stings. As they approach the entrance point of the hive, angry bees will first strike the ankles. To reduce the stings, you can use rubber bands or cord to protect your legs. You can also tuck your pants into your boots or shoes and tie your shirtsleeves with Velcro, elastic bands or wristlets. Avoid using perfumes, after-shave lotions and colognes when dealing with bees. These odors can attract suspicious bees. To avoid bee sting/irritant odors, it is important to regularly wash gloves and garments.

HIVING YOUR BEES

Before you start storing bees, it is important to decide what kind of beehive to use. There are two main types of hives: the Top-Bar and Langstroth. The Langstroth hive is widely used in the United States and is easy to manage and harvest honey from. Although there were many interesting and innovative ways to keep beehives, the most important aspect of hive architecture is the simplicity and productivity in gathering honey. Langstroth hives are portable and allow for quick honey collection. The honeycomb design is easy to make by the bees. It is rectangular in form and contains around a dozen frames where the bees can make their honey. The top and bottom are free and the combs made of wax or plastic are included. The bees form these combs and drop honey and pollen as they move. The frames will just slip out of the system, until the beekeeper is ready to extract honey and then reuse or restore it later.

The Top-Bar bee is the most common in Africa and produces honey with a higher quality. However, you cannot reuse the combs and the honey production is smaller. Amateur beekeepers love the top-bar hive, but it isn't often used in the United States. As the name suggests, the top-bar hive hangs from the top of the frame from a bar that runs across its top. This allows honeybees access to the highest point in the frame and helps them create their combs. All beekeepers who want to sell in-comb honey will find the beehive a great choice. Before you can decide on the type of beehive to use, you need to determine where and how to protect them against natural pests. Also, ensure that there aren't any local laws that prohibit you from having a beehive. It is not a good idea to go through the trouble of setting up a honey-producing beehive and then have it taken away by the police.

HOW TO DETERMINE THE KIND OF BEE YOU WANT

Entomology uses the term "race" to describe the geographical location, features and traits of the originating locale and their characteristics. There are only seven honeybee species currently recognized, with 44 subspecies. However, six to eleven species have been documented in the past. Although many bee species are similar, only Apies genus members can produce honey and store it. The dominant species in most western countries, Australia included, is the Western or European honeybee. They were originally native to Africa, Asia, Europe and Europe. Early European immigrants brought them from the New World to North America and South America.

RACE OF BEES

ITALIAN

Italian working bees have a light color while the queen is darker. This makes it easier to find her. Many workers bees have rotating markings on their stomachs.

Originating from the Italian Apennine peninsula, the original Italian bees were brought to America in 1859. They were quickly replaced by the first German or black bees by colonists.

Italian bees are the most popular bees in North America. They are sweet and successful bee breeders. They are usually raised in the south. Because they do not form large clusters like other honey bee species, they have difficulty adapting to cooler climates. Italian bees are excellent foragers and keep their hives clean and tidy. Italian bees are more inclined to swarm and have a poor sense of direction. They may move from one colony to the next and rob. It can transmit disease to the hives.

RUSSIAN

Russian bees are dark and dark brown, while the yellow portion of their abdomen is paler.

Russian bees were born in Primrosy, where Varroa and Tracheal mites also exist. We also developed strong immunity to certain hive parasites. They were imported to the United States by the U.S. Department of Agriculture in June 1997 to be resistant to mites and import into American bee stockpiles. They were made available for sale in 2000.

Russian bees are resistant to mites, and they can withstand cold environments. They are excellent overwinterers. Unfortunately, we prefer to swarm so the beekeeper must have enough space in the hive to avoid unnecessary swarming. Russian bees are especially sensitive to food availability. In environments with limited food, they can help to control brood production. Russian bees are more aggressive than other bees, but that doesn't mean they sting. Their behavior is less likely to result in them being stolen as they are more likely to engage in headbutting than stinging external threats.

CORDOVAN

Technically speaking, Cordovan bees can be described as a subclass in Italian bees that have a more vibrant hue. Both are more polite than their Italian counterparts and more likely to steal. They are fascinating to look at with their light-yellow legs, and absence of markings.

It is unclear what caused the Cordovan Bees' to take inspiration from Italian bees and create their own species of bees. This could theoretically happen with any of the bee species, but America has only found Italian bees.

CHARACTERISTICS

These bees are more kind and compassionate than the Italians they came from. They are not often included in bee boxes and are often called wilder, just like the Italians.

CAUCASIAN

The colors of the Caucasian bees range from silver-gray to deep brown. They have a longer tongue that many other bee types and can therefore take advantage of more nectar sources.

Originating from the Central Caucus region, the Caucasian bees were originally from the high valleys of the Central Caucus. This area is located between the Caspian Sea and Black seas making it particularly cold-tolerant. Propolis production is a hallmark of the Caucasian bees. Beekeepers may find it difficult to inspect the hive because of their sticky and soft propolis. They stop producing brood in the fall and prefer to overwinter well. Because they come from a cold region, they can forage even on cooler days than other bee species. The European Foul Brood is not an issue for Caucasian bees. They also seem to not swarm as much. They are not considered honeycomb producers due to their high propolis production. Instead, they prefer to keep their honey stocks. They are also susceptible to Nosema, making it difficult to find them in packets.

CARNIOLAN

Carniolan bees have black abdomens with white spots or patches. Although they are smaller than other bee species, this does not seem to affect their ability to forage and return pollen and nectar to the hive.

Carniolan bees are found in the Yugoslavia, Austrian Alps and Danube River areas. You can find them in many parts of Eastern Europe including Croatia, Bosnia-Herzegovina and Hungary.

Carniolan bees can be incredibly gentle and simple to work with. Because they are from a cooler area, they are more likely than other bee species to forage on wet and cool days. They are also great for overwintering. Even though they produce little propolis, they are able to collect their numbers quickly in the spring. Carniolan bees can also manage dearths and change brood production quickly depending on the food supply. It is slightly more likely for them to swarm than Italians so make sure they have enough space. Carniolan bees are capable of repelling rodents and hive diseases, and they also show resistance to other diseases.

BUCKFAST

Buckfast bees come in a range of colors, from yellow to brown. They resemble the honey bees that other people imagine.

The Buckfast bees are hybrids. These were created in the 20th Century by Brother Adam of Buckfast Abbey, southwest England. The American stock was imported via Canada and is now readily available.

Buckfast bees are resistant to Tracheal mites and can withstand cold temperatures. They are easy to work with, sweet and a great producer of honey. They are economical in winter markets and have a low swarming tendency. Buckfast bees thrive on long, rainy winters and are accustomed to rapidly increasing their hive size in spring.

AFRICANIZED

Africanized honeybees have a brownish, fuzzy color. They look similar to their Italian counterparts. It is difficult to tell if they are Africanized.

Chapter 15: The Beekeeping Calendar

It is important to keep track of your beekeeping schedule as a beekeeper. This will allow you to know when to inspect the hives and what other activities you can take to ensure the health of your bees. You can personalize your beekeeping experience and ensure that the bees are in good health.

You can't expect your calendar will be the same if you live somewhere like Texas. This is possible by being aware of Beekeeping Hardiness Zones. These zones are similar to those used in gardening to determine if the temperature is suitable for planting.

Beekeeping Zones

Here are some things to keep in mind when you think about Beekeeping Zones.

Zone A refers to areas in the 35-45 degree F Range. This means these areas have long, cold winters and short summers. The minimum temperatures range from 0 to 15 degrees F.

The Annual Average Temperature in Zone B is between 45 and 55 degrees F. This zone has cold winters and long summers. The minimum temperatures range from 15 to 20 degrees F.

Zone C residents will experience long, hot summers and short winters. The annual average temperature is 55-65 degrees F, with minimum temperatures ranging between 30 and 35 degrees F.

Finally, Zone D residents enjoy hot weather all year. The minimum temperature ranges from 30 to 40°F, and the average annual temperature is between 60 and 80 degrees F.

The Calendar

Here's a guide to beekeeping for the remainder of the year. It all depends on where you are located.

January

For honey shortages, Zone B and D members should inspect food reserves and feed the colony in January.

For Zones A, C, and B, be sure to check for obstructions. All others should also order new bees.

February

February is Zones A, C, and B should inspect the hive entrance for blockages.

Varroa Mites should be tested in Zones C-D. Tracheal mites should be tested and treated, and beetles and beetles must be checked. They should also begin looking for pollen replacements and conduct a comprehensive inspection.

Nosema, AFB, and EFB should be treated by those in Zone C, while those in Zone D should inspect for excess honey. Zone D residents should check for brood patterns and capped brood, as well as the location of the queen.

People in Zone B need to check their food supplies and, if necessary, do emergency feeding.

March

It's March and it's time to inspect food reserves in Zone A. Zone C should also do the same. Zones A, B, and C should provide emergency food, if necessary.

Reserve hive bodies for Zone B. Along with Zone C, zone B should start searching for the queen. They should also inspect the season for the first time.

Only Zones A, C, and B should feed pollen substitutes.

Zone D should begin looking for swarm cell and start adding honey supers and queen excluders. Zone D should start looking for brood patterns or capped brood if they exist.

Zones C andD should be taking medication for Varroa Mites and Tracheal Mites if necessary.

April

Zone B will be looking for Varroa, Nosema and Tracheal Mites in April. Zone D should begin looking for surplus honey now.

Zones C, D, and B should continue to look for supercedure cell, while Zones D and B should add honey supers and Queen Excluders and begin looking for the Queen.

Zones A through C should be looking for pollen substitutes. Zones B to C should inspect for brood and capped patterns. Also, they should conduct a comprehensive inspection of the entire season.

Zone B should also reverse beehives, while Zones A and B should inspect for eggs. Zones A, B, and C should inspect for food reserves. Zones A, B, and C should also do emergency feeding, if necessary.

May

Zone A should conduct an emergency feeding program, check for food reserves, inspect the queen and eggs, reverse hive bodies and install new hives. This is the first comprehensive inspection of the year. Zone A should also inspect for Varroa, Tracheal Mites and administer medication as necessary.

Zones A through D should now start looking for brood patterns and swarm cells.

June

Zone D should inspect for Varroa, Tracheal Mites and check honey with Zones A, B and C. In June, harvest honey with Zone B. Ventilation should be also checked.

All Zones need to start looking for swarms cells, feed pollen substitutes, add honey supers, and exclude queens.

Zones A through C should conduct a season-long inspection, while Zones B and A should set up new hives.

Zone A should also check food reserves, provide emergency feeding, check for queens and eggs, as well as reverse hive bodies.

July

Zone B should conduct emergency feedings and check food reserves in July.

Check Zones A andB for brood and capped patterns.

Zones B, D, and C should inspect for ventilation. Zones A, B, and C should inspect for excess honey. Then, harvest honey from Zones B and C!

Finally, Zone D should inspect for mites and administer medication as necessary.

August

Zones A, B, and C should prepare the hives in August for the winter season. Zones D and C should inspect the hives for beetles.

Zones A through D should be checked for mites, and if necessary, medicate.

Zones A, B and C should inspect for excess honey, while Zone B should harvest honey.

Zones C, D and E should then inspect for ventilation. All zones should also look for swarm cell activity and add queen excluders or honey supers.

Zone B should then inspect for food reserves and provide emergency feeding if necessary.

September

Zone B should provide emergency food in September. Zone C should begin looking for the queen, and eggs in September.

Zone B should check for ventilation, and then add mouse guard with zone C. Zone C should also inspect for excess honey.

Zones A through D should be inspected for mites and hives, and treated as necessary.

Finally, Zones B and A should prepare for winter.

October

In October, Zones C to A should be starting to prepare for winter. Zones B and A should also install mouse guards. Zone B should check for ventilation and Zone C should begin looking for eggs or the queen.

November

Zone A should inspect for ventilation and blockages at the entry in November. Zone C should prepare for winter now.

December

Finally, check for entry blockages in Zones A or B during December.

Chapter 16: Fungal Diseases

A beekeeper should also learn about the different fungal diseases. These are the most common fungal diseases that beekeepers should be aware of:

Chalkbrood

Because of the spores they will inhale, this type of fungus can be more dangerous to new larvae. Workers, drones, queens, and workers can all be affected by this fungus so it is important to watch out. As it eats the body, the fungus can cause the larvae. This is why the hive appears white and chalky. Laptes that have been infected will die, and it is impossible to prevent this.

Treatment & Prevention

The bees can become more ill if the honey is left to dry in their hives. As soon as you notice signs of the problem, you should do everything you can to get rid of it. Chalkbrood can be found on the entrance and bottom boards. This type of fungus is most likely to grow in spring when it is extremely wet. It is possible to stop it by ensuring that the hive has adequate ventilation.

Stonebrood

Stonebrood is another fungi that can cause mummification in honey bee colonies. This type of fungi will be most common in the soil surrounding the beehives. Because of the spores, it is difficult to detect early. Humans can also be infected with it, which could lead to respiratory problems.

The worker bees are able to combat this problem most of the time as long as their health and strength allow. The colony's ability to survive depends on the level of infection as well as the habits and hygiene of the bees.

Chapter 17: Deciding where to put your beehive

It is important to plan where your beehives will be located if you want to be a professional honey producer and beekeeper. It doesn't take a lot of space or time to manage a large number of beehives. Brilliant decisions are all that's required to make your beekeeping activities a success. These are five key considerations to make when you decide where to put your hive.

Wind direction

Neighbors

Sun exposure

Dampness

Water access

Honeybees can grow in many different places. You just need to find the right spot for your honeybees in your area. You can find beehives in cities, towns, and rural areas. A perfect spot should not be a problem for a beekeeper. You can do some research to find out what's allowed in your area. Before you quit, do a thorough analysis of all options.

Many beekeepers form groups that create communal bee yards. This allows members to realize their dream of owning beehives even though they don't live in the ideal spot. Choose a location that will not interfere with daily life. Even the most calm honey bees, which are easy to maintain, can become defensive and testy. It is not a good idea to have a beehive with sixty-thousand honey bees right next to your back door.

Beehives can be very heavy. Although they can be moved, it is best to not. You might not get the same enjoyment from the bees as your family or neighbours. Some beekeepers in the municipality have hives set up on rooftops. There are other options.

You could have hives in enclosed gardens. This would have allowed the colony flight path to be separated from the human walking path. The residents don't know that there are beehives in their area.

A few beekeepers place a line of shrubs along three wings of their beeyard. Many choose to have privacy screens outside, which require less maintenance than a living shrub. If one leaves the hive, it is possible for a shrub or bee swarm to get into your shrub. Artificial shrubs are also available

Rural dwellers have more choices. With so much open space, it is easy to choose where you want your beehive. Before you invest in honeybeekeeping, make sure to contact your local authorities.

You could get great help from other beekeepers in your area to find a better spot.

Start with just a few hives. A beekeeper who invests in too many hives at once will have more problems. If there are many beekeepers in one area, they will have more space to use.

Your beehives don't have to be placed in close proximity to wildflower fields. Many beginners are concerned about what the bees will eat, especially if they live in densely wooded areas or cities with few flowers. This should not be a problem. Honey bees will fly as far as they can to find nectar, pollen and water. The colony will be more productive if these resources are closer to the hives.

You can also plant flowers that help bees or other pollinators. Beekeeping is all about connecting with nature and understanding its balance. It is possible to plant flowers that will bloom even if there aren't any natural nectar sources.

Place your Beehives on a Stand

For easier lifting, you will need to raise your beehives from the ground. Strong wooden stands can be built and placed on top. If it works, you will find that your hive has more than one or two boxes. Are you able to lift 50 to 75 lb boxes from the top during harvest?

Your beehives can also be placed on 16-inch cement blocks. To hold them, you can place a few landscape timbers onto the blocks. Whatever method you choose, make sure your stand can hold at least a few pounds.

A simple plastic hive stand can work well for beekeepers. They are strong enough to support the hive and can also be moved easily if necessary. They also raise the beehive from the ground. This will protect your beekeeping tools against moisture from the ground.

How many colonies of bees can you keep in one place?

The foraging conditions and the climatic conditions will determine how many colonies you can keep. Talk to beekeepers near you and set realistic goals. One location may be able to provide for fifty hives during blooms, but it might prove difficult to provide enough food for twenty hives in the absence of good nectar years. This is a new perspective on honey bee management.

Bee management and health should also be considered. Safety must be taken into consideration for both bees and people. Some areas are better than others, even within your yard.

You and your bees will have a much easier time if you follow the guidelines at your site.

Accessibility for the Bees

You should position your colonies so that you can reach them in all seasons without any obstructions like snow, mud, or ice. You will need to drive or use a cart to transport a few gallons of sugar water to your bees. The more hives you have, the more feed you need. Will you be able reach a large bee colony that is heavy-weighted if it is necessary to move them? Even in snowy or muddy conditions.

Beehive Placement and Predators

You should build a fence to protect your beehives. This concern does not just pertain to animal predators. Human predators are also a concern. We hear from beekeepers who have had their beehives stolen at times. Consider a location close to your home that is easier to keep secure when you are deciding where to put your beehives. It is best to keep your colony's location secret if it is located far from people.

Put Yourself in the Right Direction

It is also important to consider the directions in which you will position your beehive. Before you place your beehive, take the time to consider all options. Ideal is a location that receives full sunlight. According to most beekeeping guidelines, beehives should face the east or southeast direction from their location. The front of the hive will receive morning sunlight, which keeps the bees active and warm. This increases the bees' motivation to get up in the morning and start their day with enthusiasm. This suggestion may not be suitable for your area. But, don't place the beehives in the cold wind direction.

Honey Bee Hives require good ventilation

When placing beehives, ventilation is an important factor. Bees that live in areas with high humidity are more susceptible to developing diseases. Flooding can be a problem if the area is too close to water sources. Remember that bees can fly and they don't necessarily need to be located right next to a stream. It is important for bees that they have easy access to water. However, be sure to place the beehives away from the floodplain and in an area with good airflow.

Take into consideration the need for a windbreak

If you live in an area with full wind, where is the best place to put your beehive? Consider providing protection for your colonies if you live in cold, windy areas. A windbreak can be a wall made of ordinary or greenery.

Shade or sun considerations:

Is it better to have your beehives in full sun or shade? Spotted sunlight is the best location for your beehives if you don't live in an area with small beetles. Bees are more active when they are in direct sunlight. Small hive beetles prefer hives that are in shade. Your colony should be located in full sunlight to prevent them from getting into your home. You must be more proactive if you only have the option of a shaded area.

These are some tips to help you choose the best place for your beehives. Don't wait until you find it. Take into consideration all these guidelines before you choose the best location for your beehives.

Chapter 18: All About The Harvest

Harvest is a time when all of the hard work has paid off. Harvest is when all the hard work of beekeeping comes to fruition and beekeepers can start selling their products. The harvest stage is a time when beekeepers will really enjoy it. If you have followed the advice in previous chapters, you'll be ready to harvest.

Harvesting honey is a very enjoyable task. Once you have harvested your first harvest, it becomes much easier. When harvesting honey, there are some things that every beekeeper needs to know (especially if you're a beginner).

It is important to understand when the harvest should take place, what tools you can use, how to filter the brood and how to keep your hive safe during harvest. It is important to know what to do with honey afterward. This chapter aims to provide you with the information you need to make the most of the honey harvesting process.

Before you harvest, make sure to inspect the brood for honey. You may need to give up some honey if there isn't. It would be tragic if your harvest was taken by bees that are starving in winter.

You should also know the blooms that occur in your area during the fall season, as well as any other factors that could affect the profits for that season. Honey is the best food for bees. Before harvesting honey, it is important to make sure they are well-fed. Honey from your hive should be used to feed your bees. There is the possibility of contamination if honey is taken from another colony.

By feeding your bees contaminated honey, you can wipe out the entire colony. Before we move on to harvesting, it is important to remember that the safety of the bees is the most important thing. Honey is an added bonus!

How do you know when it is time to harvest honey?

Honeybees are hard workers and will not give you any indications that honey is ready for harvest. So how can beekeepers know when it's available? The answer is observation. You must interrupt the bees during nectar flow to keep an eye on their progress. This should be done every two to three week. To check if they have filled the comb with nectar, you should inspect the supers.

The honey is ready to be called "Honey" when it reaches 80% moisture. You can harvest honey that has not been capped but it should still be suitable for honey-making. If the honey has not been capped, you can shake the frame and turn it upside down. You can proceed if there are not any droplets.

Take into account the botanical levels

The area surrounding the hive can be used to determine if honey is ready for harvest. You should consider the primary source for nectar in your hive: fruit trees or other plants. It is important to learn how they bloom and how long they stay in bloom. Honey harvesting season is the best time to collect honey.

The end of the nectar flow seasons is a time when bees are stocked up and you will have more honey. Spring flowers are the best time to harvest honey. A good beekeeper must ensure enough honey is available for bees in winter.

You will need the following tools to harvest honey

* This is what you will need

* Protective gear: These include beekeeping suits, gloves and a jacket, veil and boots that protect you from bee stings.

* Smoker: The smoker is important during harvest because it serves as protective gear. The smoker will be used to disperse pheromone signals and get the bees to eat honey. This will make harvesting easier.

* Use a hive tool to harvest honey.

* Bee escape: This is an optional option, but it can be used to keep the supers from your harvest. This tool must be set up at least one day before harvest.

* The bee brush: Some bees may hang onto the honeycomb while harvesting. The

bee brush can help keep them away from the capped honey. You can also keep the bees inside the hive with the bee brush while harvesting. They are soft and won't squash bees.

* You can use empty supers, or a large bucket to extract the honeycombs. The honeycombs can be placed in the bucket, or you can get supers to place the frames after you have brushed the bees off.

All tools necessary for honey harvesting can be found in your local bee shop.

How to Harvest Honey

Step 1: Examine the beehive

To harvest honey, the first step is to inspect the hive and determine the best time to harvest. The honey should not be harvested until the hives have reached 80% cap. This is when the bees will plug the combs and store the honey in the hive.

If the hive reaches 80% it means that the combs are plugged. However, honey harvested below 80% will not be allowed to be harvested and may result in poor honey production. If you collect less than 80% of the honey cap, your bees might stop producing honey completely. Therefore, it is important to inspect the honey regularly.

To get an accurate estimate, you can use a measuring tape: Measure the capped space and then compare it with the uncapped area. If it is 4 to 1, it is good.

Step 2: Look at the calendar

It is also important to take into account the harvesting seasons and times, as they vary depending on where you live. Some beekeepers harvest only once per year while others harvest three times a year in the Northern hemisphere (July August and September). It is possible to have beehives with deep double frames that can be used for winter harvest in colder climates. Don't take too much honey, as the bees will require food during winter.

Step 3: Wash your hands

Please make sure that you don't wear perfume or scented deodorant before harvesting. You don't want the bees to smell your perfume or scented deodorant. You should not wear perfume that day. If you're coming from a party, or anywhere else, it might be a good idea to take a shower.

Step four: Use your protective gear

Honeybees aren't aggressive, but opening their hives and taking out their honeycomb can trigger them. Please wear your protective gear to avoid any unpleasant situations. For this purpose, you should wear a complete beekeeping suit. This includes a long-sleeve shirt and trousers. Heavy shoes and long gloves should reach your wrists. An apiculture hat is also recommended. You should also ensure you have the right tools. If you don't want harvest to begin, it's time-wasting.

Step 5: Open the hive

You must open the hives one by one. Keep the bees safe. Smoke the smoke from the back of the hive using a smoker. To drive the bees deeper into the hive, remove the top. A smoker will prevent bees releasing a pheromone. To open the inner cover, you may need to use a small crowbar-hive tool. The bees seal it with propolis.

Sixth step: Take out the supers and frames

Once the bees have been removed, it is time to remove the frames as well as the supers from the hives. It will be easy if you only have one hive. If you have multiple hives set them 50 feet apart and cover them. To stop the bees returning, cover the frames with a blanket and close the hive.

Step seven: Cap the honeycombs

Use an uncapping knife to remove the wax-sealed honeycomb. As you extract the contents, you can use any uncapping instrument at your disposal. You can uncap honeycombs according to how they fit in the extractor. Don't overdo it, as you might lose honey.

Step 8: Take the honey out

While the honey extractor is essential, it is not mandatory for honey harvesting. You can use it after you have opened the honeycombs. Turn on the power and place the extractor over the honeycomb to extract the honey. You can still extract honey from your honeycombs without using an extractor. This is why it's not required. Some beekeepers don't use an extractor and instead place the combs on a flat surface and apply pressure.

Step 9: Strain the honey and bottle it

The final step is to strain and bottle the honey. To get rid of any foreign objects, place layers of cheesecloth over the honey. After filtration, you can bottle it. Allow the honey to settle for a few days before you use it.

Before you pour honey into the bottles or jars, wash them thoroughly. Then, replace the comb. The frames don't need to be cleaned before being returned. Please do not harvest honey from a beehive that is currently undergoing treatment. The honey may be toxic from the chemicals and antibiotics. You should treat the hive before harvesting if you have to.

Step ten: Prioritize quality over quantity

Quality is more important than the quantity of honey that you get from the harvest. Quality is more important than quantity. The quality of honey can be measured by its color, flavor, and foraging behavior.

There are over 300 honey varieties in existence around the globe. Each variety is influenced by geography, soil conditions, plant life, feeding patterns and harvest season. The National Honey Board has established the official honey color.

* White

* Amber

* Dark amber

* Light amber

* Extra-light amber

* Water-white

You can use honey's flavor to determine its quality, in addition to the above-mentioned color guidelines. You can compare notes with other beekeepers and make observations about honey harvested. If you notice anything unusual with the taste of honey, don't forget to investigate.

How often should I harvest honey

There are many factors that affect the frequency of honey harvesting. You must consider factors such as the length of winter and the number of spring blooms that bees are able to enjoy, and the temperature of the hive. It is also important to assess the vulnerability of bees to diseases and pests.

If all conditions are favorable, however, you can harvest every year. In some cases, you might harvest twice per year. You can expect exceptional yields if you have a second harvest. This will depend on the weather conditions, availability of nectar and other factors. Climate change is affecting the frequency of harvests.

Here's a tip to help you determine how often your harvest occurs: You may not harvest the first year. Instead, use the second year to track all of these factors. Find out how temperature affects honey production, quality, nectar quality, weather, and other factors. Keep track of your observations and use them to help you monitor your next harvest.

Harvesting and patience

Remember to be patient during harvest season. Every season is different. Although this book is intended to provide a guideline for standard seasons, it does not account for unusual weather conditions, which can cause seasons and weather to change. This chapter and everything in the book will show you how it "Should" look, but seasonal changes may occur. If things get delayed, you need to be patient.

You shouldn't expect to harvest honey in the first year. (I already mentioned the first-year blues). Although honey can be harvested in the first year of beekeeping, it isn't always possible. Learn more about the process and prepare to harvest the honey the second year.

This chapter is about harvesting honey. It is the most popular commodity bees produce. Although there are other products that can be obtained from bees, like wax, it is difficult for beginners to focus on the extensive harvesting process.

You will be able to learn more about how to harvest other products as you progress in your beekeeping journey. You may not need other bee products in certain areas, so find out what is available in your area before you buy it.

Honey is the first product of bees. Once you have perfected the honey-harvesting process you can move on to other products. You can harvest the honey and then start another season of beekeeping using the tips in the previous chapters.

I covered the practical aspects and history of beekeeping from Chapter 1 to this section. We will now move on to the practical sections. The next chapters will cover pest control, beekeeping tools and cleaning procedures, as well as how to avoid making mistakes. Let's start with parasites, diseases, and predators.

Conclusion

This is all you need to know about beekeeping and how to ensure that your colony will be healthy and produce as much honey for many years to come. This book will inspire you whether you're a new keeper or an experienced one.

Beekeeping is an addictive hobby. You will quickly discover that it is very addicting. You will get more honey the longer you keep your colony strong.

You will never be able to complete this hobby. Instead, you will start a cycle of caring and harvesting honey. This will last as long you want and will provide you with an organic, continuous supply of the finest honey on the planet.

You will get to know your bees and learn how to take care of them. You can do it right if you pay attention to your actions. This book will help you ensure you get the support you need to make sure you have the best possible results.

No matter how many colonies you have, this will be your first. You will fall in love not only with the quality of honey but also with the process by which it is produced.

You will see organic honey from your own backyard is the best for your health and the planet. You will be a blessing to your bees and help save a species in trouble.

You will be rewarded for your hard work if you don't mind putting in the effort. Gather the necessary supplies, plan the work, and then put it into action. You will soon have the constant supply of honey that you've been imagining.

You will see a significant improvement in your health and you will be able to share this with your family and friends.

We wish you all the best and happy beekeeping.